Rodica-Mihaela Dăneţ

A Non Standard Approach of Some Vector Lattice Concepts

AF154035

Rodica-Mihaela Dăneț

A Non Standard Approach of Some Vector Lattice Concepts

With 19 Illustrations

LAP LAMBERT Academic Publishing

Impressum / Imprint
Bibliografische Information der Deutschen Nationalbibliothek: Die Deutsche Nationalbibliothek verzeichnet diese Publikation in der Deutschen Nationalbibliografie; detaillierte bibliografische Daten sind im Internet über http://dnb.d-nb.de abrufbar.
Alle in diesem Buch genannten Marken und Produktnamen unterliegen warenzeichen-, marken- oder patentrechtlichem Schutz bzw. sind Warenzeichen oder eingetragene Warenzeichen der jeweiligen Inhaber. Die Wiedergabe von Marken, Produktnamen, Gebrauchsnamen, Handelsnamen, Warenbezeichnungen u.s.w. in diesem Werk berechtigt auch ohne besondere Kennzeichnung nicht zu der Annahme, dass solche Namen im Sinne der Warenzeichen- und Markenschutzgesetzgebung als frei zu betrachten wären und daher von jedermann benutzt werden dürften.

Bibliographic information published by the Deutsche Nationalbibliothek: The Deutsche Nationalbibliothek lists this publication in the Deutsche Nationalbibliografie; detailed bibliographic data are available in the Internet at http://dnb.d-nb.de.
Any brand names and product names mentioned in this book are subject to trademark, brand or patent protection and are trademarks or registered trademarks of their respective holders. The use of brand names, product names, common names, trade names, product descriptions etc. even without a particular marking in this works is in no way to be construed to mean that such names may be regarded as unrestricted in respect of trademark and brand protection legislation and could thus be used by anyone.

Coverbild / Cover image: www.ingimage.com

Verlag / Publisher:
LAP LAMBERT Academic Publishing
ist ein Imprint der / is a trademark of
OmniScriptum GmbH & Co. KG
Heinrich-Böcking-Str. 6-8, 66121 Saarbrücken, Deutschland / Germany
Email: info@lap-publishing.com

Herstellung: siehe letzte Seite /
Printed at: see last page
ISBN: 978-3-659-53499-7

CONTENTS

INTRODUCTION

The starting point of this work consists of several papers of the author ([D1], [D2], [D3]) published ten years ago. In these papers we gave some *algebraic descriptions* and *geometric interpretations* for some basic concepts of the vector lattice theory. This approach continued in [D5].

In this paper we begin with some facts about *ordered vector spaces* (for terminology see, for example, [C1]) and then, go step by step, discussing about *vector lattices, sublattices, lattice-subspaces, solid subsets, ideals, (o)-dense subspaces*. Then we will consider some linear operators commuting with lattice operations, such as the *Riesz homomorphisms*. We will study another type of such operators, namely the *restricted-lattice operators,* introduced in [D5]. Some *propertie*s of these operators and the problem of their *extension* will be studied. We will also introduce and study *quasi-lattice operators*.

Note that the history of *ordered vector spaces* and *vector lattices* goes back to the International Congress of Mathematicians which held at Bologna, in 1928. At this Congress, F. Riesz considered as an ordered vector space the set of all linear functionals defined on a space of real continuous functions [R].

The theory of ordered vector spaces developed with the papers of L.V. Kantorovich, written between 1935 and 1936 (see [K1] and [K2]), and independently with a paper of H. Freudenthal, published in 1936 (see [F]).

The theory has been enriched over time by many mathematicians. For example, the notion of *lattice-subspace,* used in the present work, was introduced in 1983, by I.A. Polyrakis [P1] and, independently, by S. Miyajima [M].

1

1. ORDERED VECTOR SPACES

Classical definitions

In this section we will recall classical definitions of some basic concepts in the theory of the ordered vector spaces as: order, ordered vector space, positive cone, supremum, infimum. Also we will give *algebraic descriptions* of these concepts and some *pictures*.

Definition 1.1. An *order* on a nonempty set A is a relation $"\leq"$ such that:
1) $x \leq x$, for all $x \in A$;
2) $x \leq y$ and $y \leq x$ imply that $x = y$;
3) $x \leq y$ and $y \leq z$ imply that $x \leq z$.

We also use the notation $y \geq x$ for $x \leq y$.

Definition 1.2. A vector space E, endowed with a (partial) order $"\leq"$, is called an *ordered vector space* if the order $"\leq"$ is compatible with the algebraic operations of E, that is, if:
1) $y \leq x$ in $E \Rightarrow y + z \leq x + z$ for any $z \in E$;
2) $y \leq x$ in E and $\alpha \geq 0$ in $\mathbb{R} \Rightarrow \alpha y \leq \alpha x$.

(Here E is supposed to be a *real* vector space.)

We denote by $(E, \leq)$ (or, more simply, by E) the ordered vector space E.

Definition 1.3. The set $E_+ = \{x \in E \mid 0 \leq x\}$ is called the *positive cone* in E. Its elements are called *"positive"*.

Properties of E_+. The set E_+ is a *convex cone*. This means that:
1) E_+ is a *wedge*, that is, if $x, y \in E$ and $\alpha \geq 0$, then $x + y \in E_+$ and $\alpha x \in E_+$.
2) E_+ is a *proper cone*, that is, $E_+ \cap (-E_+) = \{0\}$.

Obviously, the following equivalence holds:
$$y \leq x \Leftrightarrow x - y \in E_+.$$

2

Algebraic descriptions

Lemma 1.1. *Let* E *be an ordered vector space and* E_+ *its positive cone. Then:*

$$x \geq y \text{ in } E \Leftrightarrow x \in y + E_+ \Leftrightarrow y \in x - E_+ \Leftrightarrow -x \in -y - E_+ \Leftrightarrow -y \in -x + E_+. \quad (1)$$

Now, we accept the fact that our *intuition* is the way through which we can reach the abstract and that this way goes via the fascinating world of the *geometric interpretations*. Therefore to enter into this world, we will draw a *picture* for the geometric interpretation of the inequality $y \leq x$.

Example. $E = \mathbb{R}^2$ is an ordered vector space with the order relation "$\leq$" having the positive cone E_+ like in the next picture. For a fixed element $y \in E$, the elements $x \in E$, $y \leq x$ are exactly all elements of the set $y + E_+$.

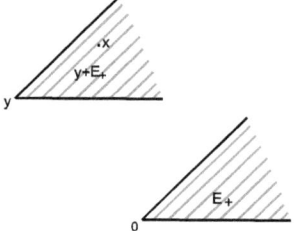

Clasical definitions

Definition 1.4. If E is an ordered vector space we say that x, y in E:

a) *have a supremum* (also called a *least upper bound*) if there exists a smallest element z in E such that $x \leq z, y \leq z$. This element z is denoted by $x \vee y$;

b) *have an infimum* (also called a *greatest lower bound*) if there exists a largest element w in E such that $w \leq x, w \leq y$. This element w is denoted by $x \wedge y$.

In other words, we have:

1) $z = x \vee y$ if and only if:
- $x \leq z, y \leq z$, and

3

- if $t \in E$ is such that $x \leq t$, $y \leq t$, then $z \leq t$;

2) $w = x \wedge y$ if and only if:

- $w \leq x$, $w \leq y$, and
- if $s \in E$ is such that $s \leq x$, $s \leq y$, then $s \leq w$.

Definition 1.5. If $A \subseteq E$ is a nonempty set we say that *there exists the supremum of* A, if there exists an element $z \in E$ such that:

1) $a \leq z$ for all $a \in A$;

2) if $a \leq t$ for all $a \in A$, then $z \leq t$.

This element is denoted by $\sup(A)$.

Definition 1.6. We say that *there exists the infimum of* the nonempty subset A of E, if there exists an element $w \in E$ such that:

1) $w \leq a$ for all $a \in A$;

2) if $s \leq a$ for all $a \in A$, then $s \leq w$.

This element is denoted by $\inf(A)$.

Algebraic descriptions

In [D1], we gave *algebraic descriptions* and a *geometric interpretation* for the elements $x \vee y$ and $x \wedge y$ (if these elements exist).

Proposition 1.2. (see [D1], and also [D2], for the proof) *Let* E *be an ordered vector space and* $x, y \in E$. *Then, for* $u, v \in E$, *the following hold:*

$$a)\ u = x \vee y \Leftrightarrow u + E_+ = (x + E_+) \cap (y + E_+); \tag{2}$$

$$b)\ v = x \wedge y \Leftrightarrow v - E_+ = (x - E_+) \cap (y - E_+). \tag{3}$$

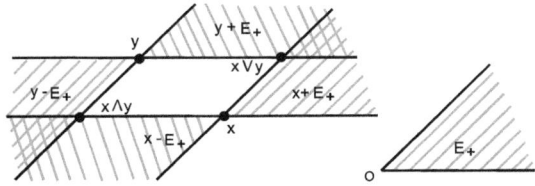

The following result is more general than Proposition 1.2.

Proposition 1.3. (see [D1]) *Let E be an ordered vector space, $A \subseteq E$ a nonempty set and consider the elements $u, v \in E$. Then:*

$$A) \ u = \sup(A) \Leftrightarrow u + E_+ = \bigcap_{x \in A}(x + E_+); \tag{4}$$

$$B) \ v = \inf(A) \Leftrightarrow v - E_+ = \bigcap_{x \in A}(x - E_+). \tag{5}$$

2. VECTOR LATTICES

Classical definitions

Definition 2.1. A *lattice* is a nonempty set A with an order $" \leq "$ such that every pair of elements $x, y \in A$ has both a supremum $x \vee y$ and an infimum $x \wedge y$.

Definition 2.2. A *vector lattice* (also called *Riesz space* or *lattice-ordered vector space*) is an ordered vector space which is also a lattice.

Remark. It is known (see, for example, [C1, Proposition 2, p.69]) that in any vector lattice E the *infinite distributivity laws* hold, that is:

a) If $(y_j)_{j \in J}$ is any family of elements of E and if $\bigvee_{j \in J} y_j$ exists, then $\bigvee_{j \in J}(x \wedge y_j)$ also exists for any $x \in E$ and the following equality holds:

$$x \wedge \left(\bigvee_{j \in J} y_j \right) = \bigvee_{j \in J}(x \wedge y_j);$$

b) If $(y_j)_{j \in J}$ is any family of elements of E and if $\bigwedge_{j \in J} y_j$ exists, then $\bigwedge_{j \in J}(x \vee y_j)$ also exists for any $x \in E$ and the following equality holds:

$$x \vee \left(\bigwedge_{j \in J} y_j \right) = \bigwedge_{j \in J}(x \vee y_j).$$

Remark. If $J = \mathbb{N}$, the formulas a) and b) are called the *countable distributivity laws*.

Suppose that E is a vector lattice.

Definition 2.3. If $x \in E$, then:

  a) the *positive part of* x is $x^+ = x \vee 0$;

  b) the *negative part of* x (which is positive!) is $x^- = (-x) \vee 0$.

  c) the *absolute value* (or, equivalently, the *modulus*) of x is
$|x| = x \vee (-x)$.

Some (basic) identities in a vector lattice

By using the *geometric interpretation* for *sup* and *inf*, the following identities can be easily intuited (see [D3]) for an arbitrary vector lattice E, $x, y, z \in E$ and $\alpha \in \mathbb{R}$.

  I) a). $(-x) \vee (-y) = -x \wedge y$.

   b). $(-x) \wedge (-y) = -x \vee y$.

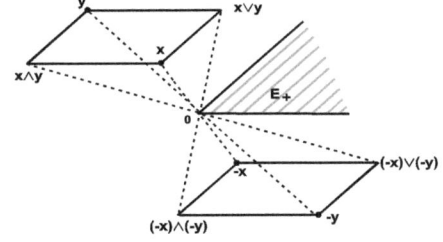

  II) a). $(\alpha x) \vee (\alpha y) = \alpha(x \vee y)$ if $\alpha > 0$.

   b). $(\alpha x) \wedge (\alpha y) = \alpha(x \wedge y)$ if $\alpha > 0$.

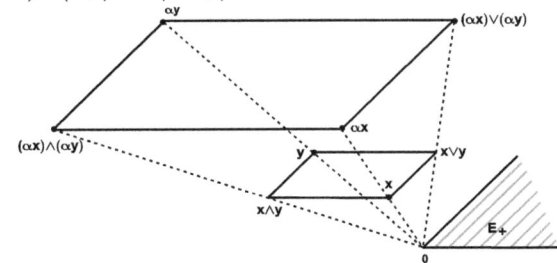

6

III). $x \vee y + x \wedge y = x + y$.

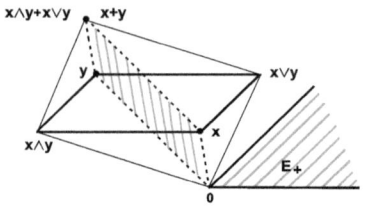

IV). $(x+y) \vee (x+z) = x + y \vee z$.

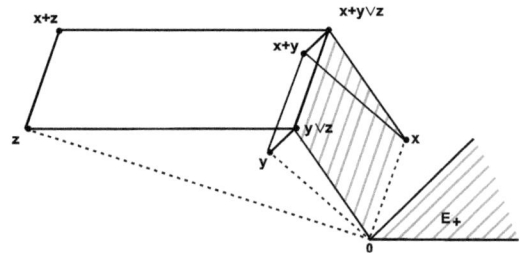

Our *proofs* for these equalities use the *algebraic descriptions* for sup and inf.

Indeed, these equalities are equivalent to the equalities between the corresponding sets. So:

I) a). The equality $(-x) \vee (-y) = -x \wedge y$ is equivalent with:

$$(-x) \vee (-y) + E_+ = -(x \wedge y - E_+) \overset{\text{P1.2}}{\Leftrightarrow}$$

$$(-x + E_+) \cap (-y + E_+) = -\left[(x - E_+) \cap (y - E_+) \right].$$

But this equality is obviously valid.
(Indeed:

$$z \in (-x + E_+) \cap (-y + E_+) \Leftrightarrow \text{ there exist } a,b \in E \text{ such that}$$

$$-x + a = z = -(y - b) \Leftrightarrow \text{ there exist } a,b \in E \text{ such that}$$

$$-(x - a) = z = -(y - b) \Leftrightarrow z \in -\left[(x - E_+) \cap (y - E_+) \right].)$$

7

I) b). Similar, the equality $(-x) \wedge (-y) = -x \vee y$ is equivalent with

$$(-x) \wedge (-y) - E_+ = -(x \vee y + E_+) \overset{\text{P1.2}}{\Leftrightarrow}$$
$$(-x - E_+) \cap (-y - E_+) = -\left[(x + E_+) \cap (y + E_+) \right]$$

and the last equality is obviously valid.

II) a). Taking into account that $E_+ = \alpha E_+$ for $\alpha > 0$ the equality

$$(\alpha x) \vee (\alpha y) = \alpha (x \vee y)$$

is equivalent with:

$$(\alpha x) \vee (\alpha y) + E_+ = \alpha (x \vee y + E_+) \overset{\text{P1.2}}{\Leftrightarrow}$$
$$(\alpha x + E_+) \cap (\alpha y + E_+) = \alpha \left((x + E_+) \cap (y + E_+) \right)$$

and the last equality is obviously valid.

II) b). The equality

$$(\alpha x) \wedge (\alpha y) = \alpha (x \wedge y)$$

is equivalent with:

$$\alpha x \wedge \alpha y - E_+ = \alpha (x \wedge y - E_+) \overset{\text{P1.2}}{\Leftrightarrow}$$
$$(\alpha x - E_+) \cap (\alpha y - E_+) = \alpha \left((x - E_+) \cap (y - E_+) \right)$$

and again the last equality is obviously valid.

III). Now to prove III), according to Proposition 1.2, we will demonstrate the following equality (between two sets):

$$x - x \vee y - E_+ = x \wedge y - E_+ - y \overset{\text{P1.2}}{\Leftrightarrow}$$
$$x - (x + E_+) \cap (y + E_+) = (x - E_+) \cap (y - E_+) - y.$$

Now we will prove the last equality.

"$\subseteq$" Let $z \in (x + E_+) \cap (y + E_+)$. Then $x - z \in x - (x + E_+) \cap (y + E_+)$ and there exist $a, b \in E_+$ such that:

$$z = \begin{cases} x + a \\ y + b \end{cases} \Rightarrow$$

$$x+y-z = \begin{cases} y-a \in y-E_+ \\ x-b \in x-E_+ \end{cases} \Rightarrow$$

$$x-z \in (x-E_+) \cap (y-E_+) - y.$$

"$\supseteq$" Conversely, let $z \in (x-E_+) \cap (y-E_+)$. Then

$$z-y \in (x-E_+) \cap (y-E_+) - y$$

and there exist $a,b \in E_+$ such that:

$$z = \begin{cases} x-a \\ y-b \end{cases} \Rightarrow$$

$$x-z+y = \begin{cases} y+a \in y+E_+ \\ x+b \in x+E_+ \end{cases} \Rightarrow$$

$$z-y = x-y-a$$

and

$$y+a = x+b \in (x+E_+) \cap (y+E_+).$$

Therefore

$$z-y \in x-(x+E_+) \cap (y+E_+).$$

IV). The equality $(x+y) \vee (x+z) = x+y \vee z$ is equivalent to the following equality:

$$(x+y) \vee (x+z) + E_+ = x+y \vee z + E_+ \overset{P1.2}{\Longleftrightarrow}$$

$$((x+y)+E_+) \cap ((x+z)+E_+) = x+((y+E_+) \cap (z+E_+)).$$

Now, we will prove the last equality.

"$\subseteq$" Let $t \in ((x+y)+E_+) \cap ((x+z)+E_+)$.

Then there exist $a,b \in E_+$ such that:

$$t = \begin{cases} x+y+a \\ x+z+b \end{cases} \Rightarrow$$

$$t-x = \begin{cases} y+a \in y+E_+ \\ z+b \in z+E_+ \end{cases}$$

Hence $t-x \in (y+E_+) \cap (z+E_+) \Rightarrow t \in x+(y+E_+) \cap (z+E_+).$

"$\supseteq$" Conversely, let $t \in x+(y+E_+) \cap (z+E_+)$. Then there exist $a,b \in E_+$ such that:

9

$$t = \begin{cases} x+y+a \in (x+y)+E_+ \\ x+z+b \in (x+z)+E_+ \end{cases} \Rightarrow$$

$$t \in \big((x+y)+E_+\big) \cap \big((x+z)+E_+\big).$$
□

Before we recall other identities in the vector lattices setting we notice something about Archimedean vector lattices (see, for example [C1, 3.1.3]).

Archimedean vector lattices

Classical definition

In what follows we will denote $\mathbb{N}^* = \mathbb{N} \setminus \{0\}$.

Definition 2.4. A vector lattice E is called *Archimedean* if for any element x, which is not negative, the set $\{\alpha x \mid 0 < \alpha \in \mathbb{R}\}$ is not bounded from above.

By using some results from [C1, 3.1.3] it follows:
Proposition 2.1. *If E is a vector lattice, then the following are equivalent:*

 i) E is an Archimedean vector lattice;

 ii) If $x > 0$ in E and $\alpha_j > 0$ in $\mathbb{R}$ $(j \in J)$ are such that $\bigwedge_{j \in J} \alpha_j = 0$, then

$\bigwedge_{j \in J} (\alpha_j x) = 0$;

 iii) If $x > 0$ in E, then $\dfrac{1}{n} x \downarrow_{n \in \mathbb{N}^*} 0$ (that is, $\bigwedge_{n \in \mathbb{N}^*} \dfrac{1}{n} x = 0$);

 iv) If $x, y \in E$ and $\alpha_n \in \mathbb{R}$ $(n \in \mathbb{N})$ and $\alpha \in \mathbb{R}$ are such that $\alpha_n x \leq y$ and $\alpha_n \to \alpha$, then $\alpha x \leq y$ (the inequality remains true if we pass to the limit);

 v) For any element $x \nleq 0$, the sequence $(nx)_{n \in \mathbb{N}}$ be not bounded from above;

 vi) Whenever $0 \leq nx \leq y$ for all $n \in \mathbb{N}, n \geq 1$ and some $y \in E_+$, then $x = 0$.

Proposition 2.2. *If E is an Archimedean vector lattice, then for any element $x > 0$ of E and any family $\left(\alpha_j\right)_{j \in J}$ of real numbers, bounded from above, there exists the element $\bigvee_{j \in J}\left(\alpha_j x\right)$ in E and the following inequality holds:*

$$\bigvee_{j \in J}\left(\alpha_j x\right) = \bigvee_{j \in J}\left(\alpha_j\right) x .$$

Remark. A similar proposition occurs if we replace the phrase "bounded from above" by "bounded from below" and the sign "$\vee$" by the sign "$\wedge$".

Algebraic descriptions

Now we will transcribe and we will prove a part of the Proposition 2.1, by using the algebraic descriptions for the inequalities and the infimums in E.

Proposition 2.3. *If E is a vector lattice, then the following are equivalent:*
    i) *E is an Archimedean vector lattice;*
    ii) *If $x \in E_+ \backslash\{0\}$ and $\alpha_j > 0$ in $\mathbb{R}$ $(j \in J)$ are such that*
$\bigcap_{j \in J}\left(\alpha_j - \mathbb{R}_+\right) = -\mathbb{R}_+$ *(that is, $\bigwedge_{j \in J} \alpha_j = 0$ in $\mathbb{R}$), then $\bigcap_{j \in J}\left(\alpha_j x - E_+\right) = -E_+$;*
    iii) *If $x, y \in E$ and $\alpha_n \in \mathbb{R} (n \in \mathbb{N})$, $\alpha \in \mathbb{R}$ are such that $y \in \alpha_n x + E_+$ for all $n \in \mathbb{N}$ and $\alpha_n \to \alpha$, then $y \in \alpha x + E_+$.*

Proof.
i) $\Rightarrow$ ii). We will prove this in two steps.
    Step 1. Let us assume that E is an Archimedean vector lattice. Firstly we will prove that

$$\bigcap_{0 < \alpha \in \mathbb{R}}\left(\alpha x - E_+\right) = -E_+ \text{ for all } x \in E_+ \backslash\{0\} . \tag{6}$$

"$\supseteq$" If $y \in -E_+$, we can write for all $\alpha > 0$:

$$y = \alpha x - (\alpha x - y) \in \alpha x - E_+ .$$

Hence, $y \in \bigcap_{0 < \alpha \in \mathbb{R}}\left(\alpha x - E_+\right)$.

"$\subseteq$" Now, let $z \in \alpha x - E_+$ for all $\alpha > 0$. If by contradiction $z \notin -E_+$, because E is an Archimedean vector lattice it follows that the set $\{\beta z \mid \beta > 0\}$ is not

11

bounded from above. But this contradicts the equality $z \in \alpha x - E_+$ for all

$\alpha > 0$ ($\Leftrightarrow \dfrac{1}{\alpha} z \in x - E_+$ for all $\alpha > 0 \Leftrightarrow \beta z \in x - E_+$ for all $\beta > 0$).

Hence, the equality (6) is true.

Step 2. Now, let $x \in E_+ \setminus \{0\}$ and $\alpha_j > 0$ $(j \in J)$ be as in the statement

"ii)", that is, such that $\underset{j \in J}{\cap} (\alpha_j - \mathbb{R}_+) = -\mathbb{R}_+$.

We have to prove that

$$\underset{j \in J}{\cap} (\alpha_j x - E_+) = -E_+. \tag{7}$$

"$\supseteq$" For $y \in -E_+$ we can write for all $j \in J$:

$$y = \alpha_j x - (\alpha_j x - y) \in \alpha_j x - E_+.$$

Hence, $y \in \underset{j \in J}{\cap} (\alpha_j x - E_+)$.

"$\subseteq$" Conversely, if $z \in \alpha_j x - E_+$ for all $j \in J$, then $z \in \alpha x - E_+$ for any

$\alpha > 0$. (Indeed, since $\underset{j \in J}{\wedge} \alpha_j = 0 < \alpha$, then there exists $j \in J$ such that $\alpha_j < \alpha$.

Therefore $z \in \alpha_j x - E_+ \subseteq \alpha x - E_+$.) It follows that $z \in \overset{(6)}{\underset{0 < \alpha \in \mathbb{R}}{\cap}} (\alpha x - E_+) = -E_+$.

Hence, $z \in -E_+$.

So, the equality (7) is valid.

ii) $\Rightarrow$ iii). We consider three cases.

Case 1. $\alpha_n \uparrow \alpha$, where α and $\alpha_n \in \mathbb{R}$ $(n \geq 1)$. (This means that $\alpha_n \leq \alpha_{n+1}$ for

all $n \in \mathbb{N}$, and $\alpha = \underset{n \in \mathbb{N}}{\vee} \alpha_n$.) It is immediate that $\underset{n \in \mathbb{N}}{\wedge} (\alpha - \alpha_n) = 0$, that is,

$\underset{n \in \mathbb{N}}{\cap} ((\alpha - \alpha_n) - \mathbb{R}_+) = -\mathbb{R}_+$, and then, according to "ii)", it follows that

$$\underset{n \in \mathbb{N}}{\cap} ((\alpha - \alpha_n) z - E_+) = -E_+ \text{ for any } z \in E_+. \tag{8}$$

If $y \in \alpha_n x + E_+$ for all $n \in \mathbb{N}$, then by considering a positive element $z \in x + E_+$

it follows that

$$\alpha x - y \in (\alpha - \alpha_n) z - E_+ \text{ for all } n \in \mathbb{N},$$

that is, $\alpha x - y \in \overset{(8)}{\underset{n \in \mathbb{N}}{\cap}} ((\alpha - \alpha_n) z - E_+) = -E_+ \Rightarrow y \in \alpha x + E_+$.

12

Case 2. $\alpha_n \downarrow \alpha$, where α and $\alpha_n \in \mathbb{R}$ $(n \geq 1)$. (This means that $\alpha_n \geq \alpha_{n+1}$ for all $n \in \mathbb{N}$, and $\alpha = \bigwedge_{n \in \mathbb{N}} \alpha_n$.) Then $-\alpha_n \uparrow -\alpha$. If $y \in \alpha_n x + E_+$, then $y \in (-\alpha_n)(-x) + E_+$. By using the *Case 1*, we obtain that $y \in (-\alpha)(-x) + E_+ = \alpha x + E_+$.

Case 3. If the sequence $(\alpha_n)_{n \in \mathbb{N}}$ with $\alpha_n \to \alpha$ is not monotonous then there exists a monotonous subsequence $\alpha_{j_n} \to \alpha$. By using one of the two previous cases, it follows that if $y \in \alpha_{j_n} x + E_+$, then $y \in \alpha x + E_+$.

iii) $\Rightarrow$ i). Let us assume iii). In particular will apply iii) for $\alpha_n = \dfrac{1}{n} \to 0$ $(n \to \infty)$. Let us remark that to prove that E is an Archimedean vector lattice, it is equivalent to prove that if $y \in \beta x + E_+$ for all $\beta > 0$, then $x \in -E_+$. But if $y \in \beta x + E_+$ for all $\beta > 0$, it follows that $\dfrac{1}{\beta}(-y) \in (-x) - E_+$ for all $\beta > 0$.

That is $-x \in \dfrac{1}{\beta}(-y) + E_+$ for all $\beta > 0$. By taking $\beta = n, n \in \mathbb{N}$, according to iii) applied for $-y$ instead of x and $-x$ instead of y, it follows that $-x \in E_+$, that is, $x \in -E_+$. □

Algebraic description and geometric interpretation for the positive part, the negative part and the absolute value in a vector lattice

By using the algebraic descriptions for sup and inf (see Proposition 1.2) it follows the following *algebraic descriptions* for the *positive part*, the *negative part* and the *absolute value* of an element $x \in E$:

$$x^+ + E_+ = (x + E_+) \cap E_+; \tag{9}$$
$$x^- + E_+ = (-x + E_+) \cap E_+; \tag{10}$$
$$|x| + E_+ = (x + E_+) \cap (-x + E_+). \tag{11}$$

The *geometric descriptions* for x^+, x^- and $|x|$ appear in the next picture:

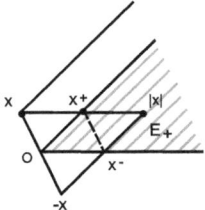

Remark the following basic properties (which can be proved by using (9), (10) and (11)):

$$x^+ \wedge x^- = 0,\qquad(12)$$
$$|x| = x^+ + x^-,\qquad(13)$$
$$|x| = x \vee (-x),\qquad(14)$$
$$x = x^+ - x^-.\qquad(15)$$

It is known (see, for example, [C1, Proposition 4, p.74]) that in order that an ordered vector space E be a vector lattice it is necessary and sufficient that for any element $x \in E$ the element x^+ should exists.

Other identities in a vector lattice

The following identities are valid in every *Archimedean vector lattice*. (It is known that every lattice identity or inequality that is true for real numbers is also true in every Archimedean vector lattice E [AB1, 8.6 Theorem]).

Notice that the following list (where $x, y, z \in E$ and $\alpha \in \mathbb{R}$) - see [AB1, p.318] - completes the list of basic identities in a vector lattice E (see the identities I)-IV)) from above:

V). $|\alpha x| = |\alpha| \cdot |x|$;

VI). $|x - y| = x \vee y - x \wedge y$;

(Compare this equality with the identity "III)", that is, with the equality $x + y = x \vee y + x \wedge y$.)

VII). $x \vee y = \dfrac{1}{2}(x + y + |x - y|)$, and

$$x \wedge y = \frac{1}{2}(x + y - |x - y|);$$

VIII). $|x| \vee |y| = \frac{1}{2}(|x + y| + |x - y|)$, and

$$|x| \wedge |y| = \frac{1}{2}(|x + y| - |x - y|);$$

IX). $|x + y| \vee |x - y| = |x| + |y|$;

X). If $\alpha > 0$, then $(\alpha x)^+ = \alpha x^+$ and $(\alpha x)^- = \alpha x^-$;

XI). (see [C1, Proposition 3, p.74]) If $\bigvee_{j \in J} x_j$ exists, then $\bigvee_{j \in J} (x_j)^+$ and

$\bigwedge_{j \in J} (x_j)^-$ exist, too, and the following equalities hold:

$$\left(\bigvee_{j \in J} x_j \right)^+ = \bigvee_{j \in J} (x_j)^+ \text{ and } \left(\bigvee_{j \in J} x_j \right)^- = \bigwedge_{j \in J} (x_j)^-.$$

Some inequalities in an Archimedean vector lattice

The following inequalities are valid in every Archimedean vector lattice E and for all $x, y, z \in E$ (see [AB1, 8.7 Corollary]):

XII). $|x + y| \le |x| + |y|$ and

$$\big||x| - |y|\big| \le |x - y|;$$

XIII). $|x \vee y - z \vee y| \le |x - z|$ and

$$|x \wedge y - z \wedge y| \le |x - z|;$$

XIV). $|x^+ - y^+| \le |x - y|$ and

$$|x^- - y^-| \le |x - y|;$$

XV). $(x + y)^+ \le x^+ + y^+$ and

$$(x + y)^- \le x^- + y^-;$$

XVI). If $x, y, z \ge 0$, then $x \wedge (y + z) \le x \wedge y + x \wedge z$.

Let us prove XVI) by using the *algebraic description* of the infimum of two elements, and the fact that every vector lattice E has the *Riesz Decomposition Property*, that is, if $0 \le t \le y + z$ in E, then there exist $a, b \in E$ such that:

$$t = a + b, \ 0 \le a \le y, \ 0 \le b \le z$$
(see, for example, [AB1, p.319]).

Equivalently, by using the algebraic description, it follows that if $t \in E_+ \cap (y + z - E_+)$, then there exist $a, b \in E_+$ with $t = a + b$ and $a \in y - E_+$, $b \in z - E_+$.

Let $x, y, z \ge 0$ in E. We have to prove XVI), that is, $x \wedge (y + z) \le x \wedge y + x \wedge z$, or, equivalently, $x \wedge (y + z) \in x \wedge y + x \wedge z - E_+$.

Let $t = x \wedge (y + z)$. It follows $t \in E_+ \cap (x - E_+)$ and $t \in E_+ \cap (y + z - E_+)$, hence $t = a + b$ with $a \in E_+ \cap (y - E_+)$ and $b \in E_+ \cap (z - E_+)$. We have $a, b \in E_+ \cap (t - E_+)$, hence $a \in (x - E_+) \cap (y - E_+)$ and $b \in (x - E_+) \cap (z - E_+)$.

Therefore
$$\begin{aligned} t = a + b &\in (x - E_+) \cap (y - E_+) + (x - E_+) \cap (z - E_+) = \\ &= (x \wedge y - E_+) + (x \wedge z - E_+) = \\ &= x \wedge y + x \wedge z - E_+. \end{aligned}$$

It follows that
$$x \wedge (y + z) \le x \wedge y + x \wedge z. \qquad \qquad \square$$

Disjoint elements

Classical definitions

Definition 2.5. (see, for example [AB1, 8.10 Definition]) Two elements x, y in a vector lattice E are *mutually disjoint*, if $|x| \wedge |y| = 0$. One writes then $x \perp y$. Notice that in [C1, Definition 2, p.77] the author used an equivalent terminology, calling $x, y \in E$ *orthogonal* if $|x| \wedge |y| = 0$.

The definition 2.5 can be extended to the subsets of E.

16

Definition 2.6. We say that a nonempty subset A of a vector lattice E is *pairwise disjoint*, if each pair of distinct elements in A is disjoint.

Definition 2.7. Two nonempty subsets A, B of a vector lattice E are said to be *disjoint* (or, equivalently, *orthogonal*) if $a \perp b$ for all $a \in A$ and $b \in B$.

The following is elementary (according to the identities (15) and (12)), but often useful: in a vector lattice E, an element x can be written as $x = x^+ - x^-$ where x^+ and x^- are disjoint elements. This decomposition of x is unique.

Proposition 2.4. ([AB1, 8.11 Theorem], [C, Proposition 1, p.73]) *If E is a vector lattice and $x = y - z$ with $y \wedge z = 0$, then $y = x^+$ and $z = x^-$, and conversely.*
Proof:
$$x^+ = x \vee 0 = (y - z) \vee 0 = y \vee z - z =$$
$$= (y + z - y \wedge z) - z =$$
$$= y.$$

Similarly $x^- = z$. □

Proposition 2.5. ([AB, 8.12 Theorem]) *If E is a vector lattice and $x, y \in E$, the following statements are equivalent:*
1). $x \perp y$;
2). $|x + y| = |x - y|$;
3). $|x + y| = |x| + |y| = |x| \vee |y|$.

Consequently, if $A = \{x_1, ..., x_n\}$ is a pairwise disjoint subset of E, then
$$\left| \sum_{i=1}^{n} x_i \right| = \sum_{i=1}^{n} |x_i| = \bigvee_{i=1}^{n} |x_i|.$$

17

A geometric interpretation for disjoint elements

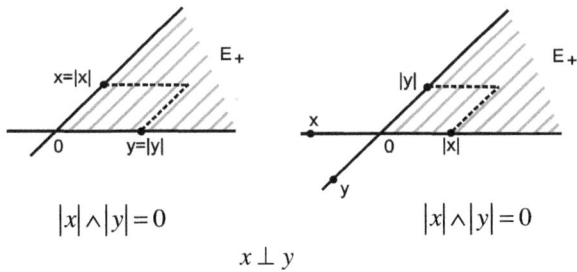

$$|x|\wedge|y|=0 \qquad\qquad |x|\wedge|y|=0$$

$$x \perp y$$

Algebraic description

Taking into account that

$$u \wedge v - E_+ = (u - E_+) \cap (v - E_+)$$

for all $u, v \in E$, we obtain the following algebraic description for disjointness.

Proposition 2.6. *Let E be a vector lattice. The elements $x, y \in E$ are disjoint if and only if*

$$\left(|x| - E_+\right) \cap \left(|y| - E_+\right) = -E_+ . \tag{16}$$

The following results can be proved by using Proposition 2.6.

Proposition 2.7. *If E is a vector lattice, $\alpha, \beta \in \mathbb{R}^*$ and x, y are two disjoint elements in E, then αx and βy are disjoint, too.*

Proof. Suppose that $x \perp y$ and $|\alpha| + |\beta| \neq 0$. Then the equality (16) from Proposition 2.6 is valid. We have to prove that $\alpha x \perp \beta y$, that is,

$$\left(|\alpha x| - E_+\right) \cap \left(|\beta y| - E_+\right) = -E_+ .$$

"$\supseteq$" Let $z \in E_+$. We can write

18

$$-z = \begin{cases} |\alpha||x| - (z + |\alpha||x|) \in |\alpha||x| - E_+ \\ |\beta||y| - (z + |\beta||y|) \in |\beta||y| - E_+ \end{cases}.$$

Hence, $-z \in (|\alpha x| - E_+) \cap (|\beta y| - E_+)$.

"$\subseteq$" Conversely, let $t \in (|\alpha x| - E_+) \cap (|\beta y| - E_+)$.

It follows that:

$$t = \begin{cases} |\alpha||x| - |\alpha|a, & \text{with } a \in E_+ \\ |\beta||y| - |\beta|b, & \text{with } b \in E_+ \end{cases} \Rightarrow$$

$$t = \begin{cases} (|\alpha| + |\beta|)|x| - (|\alpha|a + |\beta||x|) \\ (|\alpha| + |\beta|)|y| - (|\beta|b + |\alpha||y|) \end{cases}.$$

Because $|\alpha| + |\beta| \neq 0$, it follows that

$$\frac{t}{|\alpha| + |\beta|} \in \begin{cases} |x| - E_+ \\ |y| - E_+ \end{cases} \Rightarrow$$

$$\frac{t}{|\alpha| + |\beta|} \in (|x| - E_+) \cap (|y| - E_+) \overset{(16)}{=} -E_+ \Rightarrow$$

$$\Rightarrow t \in -E_+. \qquad \qquad \square$$

Proposition 2.8. (see [C1, Proposition 1, p.76]) *If E is an Archimedean vector lattice and $x, y, z \in E$ are such that $x \perp y$, $x \perp z$ and $\alpha, \beta \in \mathbb{R}$, then $x \perp (\alpha y + \beta z)$.*

Proof. Obviously, from Proposition 2.7 it follows that $x \perp y$ and $x \perp z$ implies $x \perp \alpha y$ and $x \perp \beta z$. But then, from the inequality XVI) valid in an arbitrary Archimedean vector lattice, it follows that

$$(|x| - E_+) \cap (|\alpha y + \beta z| - E_+) \overset{\text{XVI)}}{\subseteq} (|x| - E_+) \cap (|\alpha y| + |\beta z| - E_+) \subseteq$$

$$\subseteq (|x| - E_+) \cap (|\alpha y| - E_+) + (|x| - E_+) \cap (|\beta z| - E_+) = -E_+.$$

Hence, $(|x| - E_+) \cap (|\alpha y + \beta z| - E_+) \subseteq -E_+$. But obviously, the converse inclusion is valid. It follows that $(|x| - E_+) \cap (|\alpha y + \beta z| - E_+) = -E_+$, that is, $x \perp (\alpha y + \beta z)$. $\qquad \square$

Proposition 2.9. (see [C1, Proposition 2, p.76]) *If* $x \perp y$ *then:*

a) $(x+y)^+ = x^+ + y^+$;

b) $(x+y)^- = x^- + y^-$.

Proof. $x \perp y \Rightarrow |x| \wedge |y| = 0 \Rightarrow x^+ \wedge y^- = 0$.

On the other hand, it follows that $x^+ \wedge x^- = 0$. By using Proposition 2.8, we obtain

$$x^+ \wedge (x^- + y^-) = 0.$$

Similarly,

$$y^+ \wedge (x^- + y^-) = 0.$$

By using again Proposition 2.8, we infer that

$$(x^+ + y^+) \wedge (x^- + y^-) = 0.$$

From the equality

$$x + y = (x^+ + y^+) - (x^- + y^-),$$

by using Proposition 2.4, it follows that

$$(x+y)^+ = x^+ + y^+ \text{ and } (x+y)^- = x^- + y^-. \qquad \square$$

Remark. Obviously, Proposition 2.9 implies that if x and y are two *disjoint* elements in E, *then* $|x+y| = |x| + |y|$, that is, statement 3) of Proposition 2.5.

Notice that the following result is valid.

Proposition 2.10. (see, for example, [C1, Proposition 3, p.77]) *Let E be a vector lattice,* $(x_j)_{j \in J}$ *an arbitrary family of elements in E, and $y \in E$, such that $y \perp x_j$ for all $j \in J$. Then $y \perp \bigvee_{j \in J} x_j$.*

Convergence with respect to the order relation ((o)-convergence)

Classical definitions

Now, we will recall some known notions in an ordered set E (see, for example, [C1, 1.3.1]).

Definition 2.7. Let E be an ordered set. A sequence $(x_n)_n$ of elements of E is said to be *increasing* if $m \le n$ implies $x_m \le x_n$. In this case one writes $x_n \underset{n \in \mathbb{N}}{\uparrow}$ or simply $x_n \uparrow$. If, moreover, the element $x = \underset{n \in \mathbb{N}}{\vee} x_n$ exists, one writes $x_n \underset{n \in \mathbb{N}}{\uparrow} x$ or simply $x_n \uparrow x$.

Definition 2.8. Let E be an ordered set. A sequence $(x_n)_n$ of elements of E is said to be *decreasing* if $m \le n$ implies $x_m \ge x_n$. In this case one writes $x_n \underset{n \in \mathbb{N}}{\downarrow}$ or simply $x_n \downarrow$. If, moreover, the element $x = \underset{n \in \mathbb{N}}{\wedge} x_n$ exists, one writes $x_n \underset{n \in \mathbb{N}}{\downarrow} x$ or simply $x_n \downarrow x$.

Definition 2.9. (see, for example, [C1, Definition1, p.24]) The sequence $(x_n)_n$ of elements of E is said to *converge with respect to the order relation* to x (abbreviated, *(o)-converges* to x) if there exist the sequences $(a_n)_{n \in \mathbb{N}}$, $(b_n)_{n \in \mathbb{N}}$ of elements of E, such that
    1) $a_n \le x_n \le b_n$, $n \in \mathbb{N}$;
    2) $a_n \uparrow x$ and $b_n \downarrow x$.

In this case one writes $x_n \overset{o}{\to} x$. We say that x is the *limit with respect to the order relation* (in abbreviated form, the *(o)-limit*) of the sequence $(x_n)_n$ and we write $x = (o) - \lim_n x_n$.

As an example, take in the space c of all convergent real sequences - see Example 6 in Section 12 - the sequence $(x_n)_n$ with $x_n = (1,1,..., \underset{(n)}{1}, 0,...,0,...)$, $n \ge 1$, and observe that $x_n \overset{o}{\to} x$, where $x = (1,1,...)$. (Indeed taking, for example, $a_n = (1,1,..., \underset{(n-1)}{1}, 0,...,0,...)$ and $b_n = (1 + \dfrac{1}{n}, 1, 1...)$ it follows that $a_n \le x_n \le b_n$ for all $n \ge 1$, and $a_n \uparrow x$, $b_n \downarrow x$.)

Notice that the convergence with respect to the order relation can be introduced in any ordered set and consequently in any ordered vector space.

Algebraic descriptions

Proposition 2.11. *Let E be an ordered vector space. A sequence $(x_n)_n \subset E$ is:*

a) *increasing, if and only if $x_{n+1} \in x_n + E_+$ for all $n \in \mathbb{N}$;*

b) *decreasing, if and only if $x_{n+1} \in x_n - E_+$ for all $n \in \mathbb{N}$.*

Remark. If $(x_n)_n$ is increasing we have:
$$x_n \underset{n \in \mathbb{N}}{\uparrow} x \ (x \in E) \text{ if and only if } x + E_+ = \underset{n \in \mathbb{N}}{\cap}(x_n + E_+).$$
Similarly, if $(x_n)_n$ is decreasing we have:
$$x_n \underset{n \in \mathbb{N}}{\downarrow} x \ (x \in E) \text{ if and only if } x - E_+ = \underset{n \in \mathbb{N}}{\cap}(x_n - E_+).$$

Proposition 2.12. *Let E be an ordered vector space. The sequence $(x_n)_{n \in \mathbb{N}} \subset E$ (o)- converges to $x \in E$ if and only if there exist two sequences in E, $(a_n)_{n \in \mathbb{N}}$ and $(b_n)_{n \in \mathbb{N}}$, such that $a_{n+1} \in a_n + E_+$, $b_{n+1} \in b_n - E_+$, $x_n \in (a_n + E_+) \cap (b_n - E_+)$ for all $n \in \mathbb{N}$, and, moreover, $x + E_+ = \underset{n \in \mathbb{N}}{\cap}(a_n + E_+)$ and $x - E_+ = \underset{n \in \mathbb{N}}{\cap}(b_n - E_+)$.*

Proposition 2.13. ([C1, Proposition 1, p.24] and [AB1, 8.15 Lemma]) *The convergence with respect to the order relation of a vector lattice E has the following properties:*

1). *If $x_n \uparrow$ (or $x_n \downarrow$), then $x_n \overset{o}{\to} x$ if and only if $x_n \uparrow x$ ($x_n \downarrow x$, respectively).*

2). *Any (o)-convergent sequence $(x_n)_{n \in \mathbb{N}}$ is bounded (that is, there exist $a, b \in E$ such that $x_n \in (a + E_+) \cap (b - E_+)$ for all $n \in \mathbb{N}$).*

3). *If $y_n \in x_n + E_+$, and $x_n \overset{o}{\to} x$, $y_n \overset{o}{\to} y$, then $y \in x + E_+$.*

4). If $x_n \xrightarrow{o} x$ and $x_n \xrightarrow{o} y$, then $x = y$.

5). If $x_n \xrightarrow{o} x$ and $y_n \xrightarrow{o} y$, then $x_n + y_n \xrightarrow{o} x + y$.

6). If for all $n \in \mathbb{N}$, $x_n \in (y_n + E_+) \cap (z_n - E_+)$ and $y_n \xrightarrow{o} x$, $z_n \xrightarrow{o} x$, then $x_n \xrightarrow{o} x$.

7). If $x_n \xrightarrow{o} x$, $y_n \xrightarrow{o} y$, then $x_n \vee y_n \xrightarrow{o} x \vee y$ and $x_n \wedge y_n \xrightarrow{o} x \wedge y$.

8). If $x_n \xrightarrow{o} x$, then $x_n^+ \xrightarrow{o} x^+$, $x_n^- \xrightarrow{o} x^-$ and $|x_n| \xrightarrow{o} |x|$.

9). If $x_n \xrightarrow{o} x$, then $\alpha x_n \xrightarrow{o} \alpha x$ for each $\alpha \in \mathbb{R}$.

Proof.

1). Let $(x_n)_n \subset E$ be such that $x_n \uparrow$ and $x_n \xrightarrow{o} x$ and let $(a_n)_n$ and $(b_n)_n$ like in Definition 2.9. Because $x_m \in b_n - E_+$ for any m and n, it follows that $x_m \in x - E_+$ for all $m \in \mathbb{N}$. Suppose that for an element $y \in E$ we have $x_m \in y - E_+$ for all $m \in \mathbb{N}$. Then $a_m \in y - E_+$ for all $m \in \mathbb{N}$, whence $x \in y - E_+$. It follows that $x = \bigvee_{n \in \mathbb{N}} x_n$. Conversely, if $x_n \uparrow x$, then by taking $a_n = x_n$ and $b_n = x$, we obtain that the conditions from Proposition 2.12 are satisfied and, therefore, $x_n \xrightarrow{o} x$. The case of the decreasing sequence $(x_n)_{n \in \mathbb{N}}$ may be similarly treated.

2). Obviously from the conditions of Proposition 2.12, it follows that $x_n \in (a_1 + E_+) \cap (b_1 - E_+)$ for all $n \in \mathbb{N}$.

3). Consider $(a_n)_{n \in \mathbb{N}}$ and $(b_n)_{n \in \mathbb{N}}$ such that $a_n \uparrow x$, $b_n \downarrow y$ and, for all $n \in \mathbb{N}$, $a_n \in x_n - E_+$, $b_n \in y_n + E_+$. From the hypothesis we have also that $y_n \in x_n + E_+$, for all $n \in \mathbb{N}$. It follows that, $b_m \in a_n + E_+$ for all $m, n \in \mathbb{N}$. Then it easily follows that $y \in x + E_+$.

4). This is an immediate consequence of 3).

5). Beacause $x_n \xrightarrow{o} x$ and $y_n \xrightarrow{o} y$, there exist $(a_n)_n$, $(b_n)_n$, $(c_n)_n$, $(d_n)_n$ such that:

a). $a_n \uparrow x$, $b_n \downarrow x$ and, for all $n \in \mathbb{N}$, $x_n \in (a_n + E_+) \cap (b_n - E_+)$;

b). $c_n \uparrow y$, $d_n \downarrow y$ and, for all $n \in \mathbb{N}$, $y_n \in (c_n + E_+) \cap (d_n - E_+)$.

It is immediate that: $(a_n + c_n) \uparrow (x + y)$, $(b_n + d_n) \downarrow (x + y)$ and $x_n + y_n \in$ $(a_n + b_n + E_+) \cap (c_n + d_n - E_+)$, that is, $x_n + y_n \overset{o}{\to} x + y$.

6). Since $y_n \overset{o}{\to} x$ there exists $(a_n)_{n \in \mathbb{N}}$ such that $a_n \uparrow x$ and $y_n \in a_n + E_+$ for all $n \in \mathbb{N}$. There also exists $(b_n)_{n \in \mathbb{N}}$ such that $b_n \downarrow x$ and $z_n \in b_n - E_+$ for all $n \in \mathbb{N}$. Since $a_n \in y_n - E_+$, $y_n \in x_n - E_+$, $x_n \in z_n - E_+$ and $z_n \in b_n - E_+$ for all $n \in \mathbb{N}$ it follows that $a_n \in x_n - E_+$ and $b_n \in x_n + E_+$ for all $n \in \mathbb{N}$. Then $x_n \overset{o}{\to} x$.

7). Let $(a_n)_n$ and $(b_n)_n$ be two sequences such that $a_n \uparrow$, $b_n \downarrow$ and $x_n \in (a_n + E_+) \cap (b_n - E_+)$ for all $n \in \mathbb{N}$. Let also $(c_n)_n$ and $(d_n)_n$ be two sequences such that $c_n \uparrow$, $d_n \downarrow$ and $y_n \in (c_n + E_+) \cap (d_n - E_+)$ for all $n \in \mathbb{N}$. We obviously have

$$x_n \vee y_n \in (a_n \vee c_n + E_+) \cap (b_n \vee d_n - E_+)$$

for all $n \in \mathbb{N}$. By taking into account that the sequences $(a_n)_n$ and $(c_n)_n$ are increasing we get

$$x \vee y = \left(\bigvee_{m \in \mathbb{N}} a_m \right) \vee \left(\bigvee_{n \in \mathbb{N}} c_n \right) = \bigvee_{m, n \in \mathbb{N}} (a_m \vee c_n) = \bigvee_{n \in \mathbb{N}} (a_n \vee c_n),$$

hence

$$a_n \vee c_n \uparrow x \vee y. \tag{17}$$

Now we will use that in any vector lattice the infinite distributivity laws hold (see the Remark at the beginning of the Section 2 in this paper and [C1, Proposition 2, p.69]). In particular, the countably distributivity laws hold. Then, in a similar way with the proof of (17), by taking into account that the sequences $(b_n)_n$ and $(d_n)_n$ are decreasing we get:

$$x \vee y = \left(\bigwedge_{m \in \mathbb{N}} b_m \right) \vee \left(\bigwedge_{n \in \mathbb{N}} d_n \right) = \bigwedge_{m, n \in \mathbb{N}} (b_m \vee d_n) = \bigwedge_{n \in \mathbb{N}} (b_n \vee d_n),$$

hence

$$b_n \vee d_n \downarrow x \vee y. \tag{18}$$

As a consequence of (17) and (18) we obtain that $x_n \vee y_n \overset{o}{\to} x \vee y$.

In the same manner we can show that $x_n \wedge y_n \overset{o}{\to} x \wedge y$.

8). This property follows from "7)".

9). It is immediate (for $\alpha \neq 0$ we have two cases: $\alpha > 0$ and, then, $\alpha < 0$). $\quad \square$

3. SUBLATTICES

Classical definition

Let E be a vector lattice and $G \subseteq E$ a vector subspace.

Definition 3.1. We say that G is a *sublattice* of E if for all $u, v \in G$, there exists $u \vee v$ in E and $u \vee v \in G$ (or, equivalently, for all $u, v \in G$, there exists $u \wedge v$ in E and $u \wedge v \in G$).

Algebraic description

The following result gives an algebraic description of the concept of *sublattice*.

Proposition 3.1. *Let E be a vector lattice and $G \subseteq E$ a vector subspace. Then, G is a sublattice if and only if for all $u, v \in G$, there exists $z \in G$ such that $z + E_+ = (u + E_+) \cap (v + E_+)$ and, in this case, $z = u \vee v$.*

Proof.

"$\Rightarrow$" Take $z = u \vee v$ (calculated in E!) and apply the identity (2) (see Proposition 1.2).

"$\Leftarrow$" Let $u, v \in G$. We know that:

$$\text{there exists } z \in G \text{ such that } z + E_+ = (u + E_+) \cap (v + E_+) \tag{19}$$

$$\text{and } u \vee v + E_+ = (u + E_+) \cap (v + E_+). \tag{20}$$

We have to prove that $z = u \vee v$.

But $z \in z + E_+ \overset{(19)}{=} (u + E_+) \cap (v + E_+) \Rightarrow z \geq u, \; z \geq v$.

Now let $t \in E$ such that $t \geq u, t \geq v \Rightarrow t \in (u + E_+) \cap (v + E_+) \overset{(19)}{=} z + E_+ \Rightarrow t \geq z$.

Otherwise, from (19) and (20) we deduce that $z + E_+ = u \vee v + E_+$. Then $z \in u \vee v + E_+$ and $u \vee v \in z + E_+$, that is, $u \vee v \leq z \leq u \vee v$. Hence $z = u \vee v$.

25

Hence $z = u \vee v$. But $z \in G$ and therefore $u \vee v \in G$. □

Proposition 3.2. (see [D3, Proposition 1]) *Let E be a vector lattice and $G \subseteq E$ an ordered (vector) subspace (-see Definition 4.1 below). Then G is a sublattice of E if and only if for all $u, v \in G$, the following equality holds:*

$$u \vee v + G_+ = (u + E_+) \cap (v + E_+) \cap G.$$

Proof.

" $\Rightarrow$ " " $\subseteq$ " Take $a \in G_+$ and $u, v \in G$ (hence $u \vee v + a \in u \vee v + G_+$).

Because G is a sublattice, it follows:

$$u \vee v + a \in (u \vee v + G_+) \cap G \subseteq (u \vee v + E_+) \cap G \overset{(19)}{=}$$
$$= (u + E_+) \cap (v + E_+) \cap G.$$

" $\supseteq$ " Conversely, let $w \in (u + E_+) \cap (v + E_+) \cap G \overset{(19)}{\Rightarrow} w \in G$ and $w \in u \vee v + E_+$

$\Rightarrow$ there exists $s \in E_+$ such that $w = u \vee v + s$. $\qquad$ (*)

But $w \in G$ and $u \vee v \in G \overset{(*)}{\Rightarrow} s \in G \cap E_+ = G_+ \Rightarrow w = u \vee v + s \in u \vee v + G_+$.

" $\Leftarrow$ " $u \vee v = u \vee v + 0 \in u \vee v + G_+ = (u + E_+) \cap (v + E_+) \cap G =$
$$\overset{(19)}{=} (u \vee v + E_+) \cap G \Rightarrow u \vee v \in G \Rightarrow G \text{ sublattice.} \qquad □$$

The following result is a consequence of Proposition 3.1.

Corollary 3.3. *If E is a vector lattice and $G_1, G_2 \subseteq E$ are two sublattices, then $G_1 \cap G_2$ is a sublattice, too.*

Proof. Let $u, v \in G_1 \cap G_2$. According to Proposition 3.1, there exist $z_1 \in G_1$ and $z_2 \in G_2$ such that

$$z_1 + E_+ = (u + E_+) \cap (v + E_+),$$
$$z_2 + E_+ = (u + E_+) \cap (v + E_+).$$

It follows that $z_1 + E_+ = z_2 + E_+$. So we obtain $z_1 \in z_1 + E_+ = z_2 + E_+$ and hence it follows $z_1 \geq z_2$. The converse inequality ($z_2 \geq z_1$) is similar. Therefore $z_1 = z_2 \overset{denoted}{=} z$. It follows that there exists $z \in G_1 \cap G_2$ such that

$$z + E_+ = (u + E_+) \cap (v + E_+).$$

Then by applying again Proposition 3.1, it follows that $G_1 \cap G_2$ is a sublattice of E. $\qquad\qquad\qquad\qquad\qquad\qquad\qquad\qquad\qquad\qquad\qquad\qquad\qquad\quad$ $\square$

4. LATTICE-SUBSPACES

Classical definitions

Let E be an ordered vector space and $G \subseteq E$ a vector subspace.
Definition 4.1. We say that G is an *ordered subspace* of E if G is ordered by the induced ordering (that is by the cone $G_+ = G \cap E$).

Now, suppose that G is an ordered subspace of E.
Definition 4.2. G is called a *lattice-subspace* of E (see [P1]) if G is a vector lattice, that is, for each $u, v \in G$, the supremum $u \vee_G v$ of $\{u,v\}$ (calculated in G) exists in G. (Mention that $u \vee_G v$ is also denoted by $\sup_G \{u,v\}$ or $u \nabla v$.)

But what means $u \vee_G v = z$?
$$u \vee_G v = z \Leftrightarrow \text{ 1) } z \in G \text{ and } u \le z, v \le z, \text{ and}$$
$$\text{2) for each } t \in G, \text{ with } u \le t, v \le t, \text{ it follows } z \le t.$$

It is clear that
$$u \vee v \le u \vee_G v$$
whenever the supremum $u \vee v$ of $\{u,v\}$ exists in E.

Similarly we will denote by $u \wedge_G v$ the infimum of $\{u, v\}$ (calculated in G), if this element exists in G. Hence, we have:
$$u \wedge_G v = w \Leftrightarrow \text{ 1') } w \in G \text{ and } w \le u, w \le v, \text{ and}$$
$$\text{2') for each } s \in G, \text{ with } s \le u, s \le v, \text{ it follows } s \le w.$$

Obviously, G is a *lattice-subspace* of E if and only if for each $u, v \in G$ the infimum $u \wedge_G v$ of $\{u, v\}$ exists in G.

27

It is clear that

$$u \wedge_G v \le u \wedge v$$

whenever the infimum $u \wedge v$ of $\{u,v\}$ exists in E.

Remark that we can replace $\{u,v\}$ with a nonempty subset $A \subseteq G$ supposing that there exist the following elements

$$z = \sup_G A \text{ and } t = \inf_G A.$$

Algebraic description (see [D3])

Similar with the algebraic identities (2) - (5) for sup and inf calculated in the whole space E, we have:

$$u \vee_G v + G_+ = (u + G_+) \cap (v + G_+); \tag{21}$$

$$u \wedge_G v - G_+ = (u - G_+) \cap (v - G_+); \tag{22}$$

$$\sup_G A + G_+ = \bigcap_{x \in A} (x + G_+); \tag{23}$$

$$\inf_G A - G_+ = \bigcap_{x \in A} (x - G_+). \tag{24}$$

Note that if E is a vector lattice and $u \vee_G v = u \vee v$ for any $u, v \in G$ ($G \subseteq E$ an ordered subspace) then G is a sublattice of E (see [P1]).

Note that $u \vee_G v$ depends on the subspace $G \subseteq X$ ($u, v \in G$). In other words, in this kind of subspaces we have the induced ordering and a lattice structure but not the induced one (see [P4]).

From the identities (2), (3) (see Proposition 1.2), and (21), (22) it follows immediate that if $G \subseteq E$ is a lattice-subspace and $u, v \in G$ are such that there exist $u \vee v$ and $u \wedge v$, calculated in E (what's happening if, for example, E is a vector lattice) then we have:

$$u \wedge_G v \le u \wedge v \le u \vee v \le u \vee_G v.$$

Remark. *The class of all lattice-subspaces in a vector lattice E is larger then that of all sublattices in E, because any sublattice is a lattice-subspace but the*

28

converse is not true in general.

The following algebraic description clarifies more the difference between a lattice-subspace and a vector sublattice in a vector lattice E.

Proposition 4.1. (see Proposition 2 in [D3]) *Let E be a vector lattice and $G \subseteq E$ an ordered vector subspace. Then G is a lattice-subspace if and only if for all $u, v \in G$, there exists $z \in G$ such that*

$$z + G_+ = (u + G_+) \cap (v + G_+) \qquad (*)$$

(and, in this case $z = u \vee_G v$).

Proof. G is a lattice-subspace $\Leftrightarrow$ for all $u, v \in G$, there exists $u \vee_G v \in G$.

Denote $z = u \vee_G v \overset{(19)}{\Leftrightarrow} z + G_+ = (u + G_+) \cap (v + G_+)$. $\qquad\qquad \square$

To compare the notion of *lattice-subspace* with the notion of *sublattice*, recall Proposition 3.1: G *is a sublattice in E if and only if for all $u, v \in G$, there exists $z \in G$ such that*

$$z + E_+ = (u + E_+) \cap (v + E_+) \qquad (**)$$

Note that with Proposition 3.1 and Proposition 4.1 the previous Remark is obvious.

Remark. Another *consequence* of Proposition 4.1 is that, unlike the case of sublattices, *the intersection of two lattice-subspaces can be not a lattice-subspace.*

A brief history

In the papers of I.A. Polyrakis we find a *short history of the lattice-subspace notion (*for example in [P3, 1999]).

The *notion of lattice-subspace* was introduced by Polyrakis in 1983 - see [P1]. In this paper it is proved that each infinite dimensional closed lattice-subspace of l_1 is order-isomorphic to l_1.

At the same time, the *notion of lattice-subspace* was introduced *independently*, by S. Miyajima in 1983 - see [M], where the term "*quasi-sublattice*" is used and it is proved that G is a lattice-subspace of the vector lattice E, if and only if G is the range of a positive projection from the sublattice $S(G)$ (generated by G) onto G.

Lattice-subspaces appear in the work of many authors, in their attempt to study the subspaces of a vector lattice E which are the range of a positive projection P, that is $G = P(E)$. Then it easy to show that G is a lattice-subspace of E, with $u \vee_G v = P(u \vee v)$ for all $u, v \in G$ but as it is remarked in [AAP, 1994], there are finite-dimensional lattice-subspaces which are not the range of a positive projection.

In 1992, C.D. Aliprantis and D. Brown understood the meaning of the lattice-subspaces in economics and posed the problem of *the study of finite-dimensional lattice-subspace*. This problem is interesting even in $\mathbb{R}^n$, because many economic models, as the famous *Arrow-Debreu model*, are finite. This problem was the motivation for [AAP], where the lattice-subspaces of $\mathbb{R}^n$ are studied (see also [P2]).

5. SOLID SUBSETS. IDEALS

Classical definitions (see, for example, [C1, p.91])

Definition 5.1. Let E be a real vector lattice and for each $x \in E$, consider the set $Z(x) = \{y \in E \mid |y| \leq |x|\}$ and call it the *solid kernel* of $\{x\}$.

Definition 5.2. We say that a (nonempty) set $A \subseteq E$ is *solid* if for each $x \in A$, it follows that $Z(x) \subseteq A$.

Definition 5.3. For an arbitrary (nonempty) set $B \subseteq E$, we define the set $so(B) = \bigcup_{x \in B} Z(x)$ and it call the *solid hull* of B.

Definition 5.4. A vector subspace G of a vector lattice E is called an *ideal* (or, equivalently, a *normal subspace*) if G is a solid subset of E.

Lemma 5.1. *If* $x, y, z \in E$ *(* E *a vector lattice) are such that* $|z| \le |x + y|$ *then there exist* $u, v \in E$ *with* $z = u + v$ *and* $|u| \le |x|, |v| \le |y|$.

Proof. Because E is a vector lattice, it has the *Riesz Decomposition Property* (that is if $a, b, c \in E_+$ are such that $0 \le a \le b + c$ then there exist $r, s \in E$ such that $a = r + s$ and $0 \le r \le b, 0 \le s \le c$).
Then, because $0 \le z^+ \le |x| + |y|$ there exist $m, n \in E$ such that
$$z^+ = m + n, 0 \le m \le |x|, 0 \le n \le |y|.$$
Similar, there exist $p, q \in E$ such that
$$z^- = p + q, 0 \le p \le |x|, 0 \le q \le |y|.$$
Then, it follows:
$$z = z^+ - z^- = (m - p) + (n - q).$$
Denote $u = m - p$ and $v = n - q$.
We have $z = u + v$ and $|u| \le |x|, |v| \le |y|$. (Indeed, for example, from $0 \le m \le |x|$ and $-|x| \le -p \le 0$ it follows $-|x| \le m - p \le |x|$ that is $|m - p| \le |x|$). $\qquad \square$

Proposition 5.2. (properties of the sets $Z(x)$)
      a) $Z(x + y) \subseteq Z(x) + Z(y)$ *for all* $x, y \in E$.
      b) $Z(\alpha x) = |\alpha| Z(x)$ *for all* $x \in E$ *and* $\alpha \in \mathbb{R}$.

Proof. a) Let $z \in Z(x + y) \Rightarrow |z| \le |x + y|$. With the previous lemma, there exist $u, v \in E$ such that
$$z = u + v \text{ and } |u| \le |x|, |v| \le |y| \Rightarrow z \in Z(x) + Z(y).$$
b) Obviously the equality is valid for $\alpha = 0$, because $Z(0) = \{0\}$. Now we assume $\alpha \ne 0$. Then

$$Z(\alpha x) = \left\{ y \in E \,\middle|\, |y| \le |\alpha x| \right\} =$$

$$= \left\{ y \in E \,\middle|\, \left| \frac{1}{\alpha} y \right| \le |x| \right\} =$$

$$= |\alpha| \left\{ \frac{1}{|\alpha|} y \in E \,\middle|\, \left| \frac{1}{|\alpha|} y \right| \le |x| \right\}$$

$$= |\alpha| Z(x). \qquad \square$$

The following result is immediate.

Proposition 5.3. *A set $A \subseteq E$ (E a vector lattice) is a solid set if and only if $Z(x) \subseteq A$ for all $x \in A$.*

Remark. With the Lemma 5.1 it is immediate that if $A, B \subseteq E$ are two solid sets then their sum $A + B$ is a solid set, too.

Algebraic descriptions

Firstly, we establish an algebraic description for the inequality $|y| \le |x|$.

Lemma 5.4. *Let $x, y \in E$. Then $|y| \le |x|$ if and only if*

$$(x + E_+) \cap (-x + E_+) \subseteq (y + E_+) \cap (-y + E_+).$$

Proof.

$$|y| \le |x| \Leftrightarrow |x| \in |y| + E_+ \Leftrightarrow$$

$$\Leftrightarrow |x| + E_+ \subseteq |y| + \underbrace{E_+ + E_+}_{\substack{E_+ \\ (E_+ \text{ is a cone})}} \Leftrightarrow$$

$$\Leftrightarrow |x| + E_+ \subseteq |y| + E_+ \Leftrightarrow$$

$$\underset{(11)}{\Leftrightarrow} (x + E_+) \cap (-x + E_+) \subseteq (y + E_+) \cap (-y + E_+). \quad \square$$

Proposition 5.5.

　　a) A set $A \subseteq E$ is a solid set if and only if for each $x \in A$ and $y \in E$

with

$$(x+E_+)\cap(-x+E_+)\subseteq(y+E_+)\cap(-y+E_+)$$

it follows that $y\in A$.

 b) A vector subspace $G\subseteq E$ is an ideal if and only if for each $x\in G$ and $y\in E$ with

$$(x+E_+)\cap(-x+E_+)\subseteq(y+E_+)\cap(-y+E_+)$$

it follows that $y\in G$.

 c) Any ideal $G\subseteq E$ is a sublattice (but conversely is not true).

We notice that to prove "c)" we apply the previous lemma for $y=|x|$.

Remark. With the Lemma 5.4 we can prove for example Proposition 5.2 b), that is, the equality

$$Z(\alpha x)=|\alpha|Z(x),\text{ for all }x\in E\text{ and }\alpha\in\mathbb{R}^*$$

(for $\alpha=0$, the equality is obviously valid).

Indeed, we have:

$$y\in Z(\alpha x)\Leftrightarrow|y|\leq|\alpha x|\Leftrightarrow$$

$$(\alpha x+E_+)\cap(-\alpha x+E_+)\subseteq(y+E_+)\cap(-y+E_+)\overset{\alpha>0}{\Leftrightarrow}$$

$$(x+E_+)\cap(-x+E_+)\subseteq\left(\frac{1}{|\alpha|}y+E_+\right)\cap\left(-\frac{1}{|\alpha|}y+E_+\right)\Leftrightarrow$$

$$\frac{1}{|\alpha|}y\in Z(x)\Leftrightarrow y\in|\alpha|Z(x). \qquad\qquad\Box$$

We know that a sublattice need not be an ideal. For instance (see, for example, [AB1, p. 321]) $C([0,1])$ is a sublattice of $\mathbb{R}^{[0,1]}$, but is not an ideal. On the other hand, the l_p- spaces are ideals in $\mathbb{R}^\mathbb{N}$ - see Examples 8 and 9 in Section 12.

It is easy to prove that the intersection of any two sublattices (ideals) is a sublattice (an ideal, respectively). But the situation for sums is more complicated. So, the sum of two ideals of a vector lattice is an ideal, too - see

[M-N, Proposition 1.2.2]. But the sum of two sublattices needs not to be a sublattice. For example, $U_1 = \{a\mathbf{1} \mid a \in \mathbb{R}\}$ and $U_2 = \{aj \mid a \in \mathbb{R}\}$ are sublattice in $C([0,1])$, where $j(t) = t$ for all $t \in [0,1]$, but $U_1 + U_2$ fails to be a sublattice of $C([0,1])$ - see [M-N, p.12]. This could be another reason for the fact that a vector sublattice may not be an ideal.

The following result gives a necessary and sufficient condition such that an ideal is a vector sublattice.

Proposition 5.6. (see [AB1, 8.13 Theorem]) *A sublattice G of a vector lattice E is an ideal if and only if x is in E, $0 \le x \le y$ and $y \in G$ imply $x \in G$ (or equivalently, $x \in E_+$ and $y \in (x + E_+) \cap G$ imply $x \in G$).*

Proof. It must to prove only the "if" part. So, assume that:
$$x \in E_+ \text{ and } y \in (x + E_+) \cap G, \text{ imply } x \in G. \tag{*}$$
Now let $u \in E$ and $v \in G$ such that
$$(v + E_+) \cap (-v + E_+) \subseteq (u + E_+) \cap (-u + E_+).$$
Since G is a sublattice it follows that $|v| \in G$. But we observe that:
$$(u + E_+) \cap (-u + E_+) \subseteq (u + E_+) \cap E_+.$$
(Indeed, if $z \in (u + E_+) \cap (-u + E_+) \Rightarrow$ there exist $a, b \in E_+$ such that
$$z = \begin{cases} u + a \\ -u + b \end{cases} \Rightarrow 2z = a + b \in E_+ \Rightarrow z \in E_+;$$
so, $z \in (u + E_+) \cap E_+$.)
Therefore, it follows that
$$|v| + E_+ = (v + E_+) \cap (-v + E_+) \subseteq (u + E_+) \cap E_+ = u^+ + E_+$$
and hence $u^+ \in G$ because $|v| \in G$ and we apply (*) for $x = u^+$ and $y = |v|$. Similar we can prove that $u^- \in G$. Hence $u = u^+ - u^- \in G$, so G is an ideal. $\square$

Now by using Lemma 5.4, we will prove a known property concerning solid sets in a vector lattice.

Proposition 5.7. *Let* $(A_\delta)_{\delta \in \Delta}$ *be a family of solid sets in a vector lattice* E. *Then, the intersection* $\bigcap_{\delta \in \Delta} A_\delta$ *is a solid set, too.*

Proof. Obviously, we can assume that the previous intersection is nonempty. Let $x \in \bigcap_{\delta \in \Delta} A_\delta$ and $y \in E$ with $(x + E_+) \cap (-x + E_+) \subseteq (y + E_+) \cap (-y + E_+)$. Then, because for each $\delta \in \Delta$, $x \in A_\delta$ and A_δ is a solid set, it follows that $y \in A_\delta$. $\quad\quad\quad\square$

Remark. The previous result shows us that the intersection of a family of ideals is an ideal, too. Therefore we consider:

        a) the *solid hull of a nonempty set* A, denoted by $so(A)$, and defined as $\bigcap_{\substack{B \supseteq A \\ B \subseteq E \\ solid}} B$;

        b) the *ideal generated by a set* A, denoted by $I(A)$, and defined as $\bigcap_{\substack{G \supseteq A \\ G \subseteq E \\ ideal}} G$;

        c) the *principal ideal generated by an element* $x \in E$, denoted by E_x and defined as a particular case of "b)", taking $A = \{x\}$.

The following three results give descriptions for $so(A)$, $I(A)$ and E_x, respectively.

Proposition 5.8. *For a set* $A \subseteq E$ *(* E *vector lattice)* $so(A) = \bigcup_{x \in A} Z(x)$.

Proof. Firstly we remark that $\bigcup_{x \in A} Z(x)$ is a solid set that contains A. (Indeed if $z \in \bigcup_{x \in A} Z(x)$ and $y \in E$ is such that $|y| \le |z|$ it follows that there exist $x \in A$ such that $z \in Z(x)$, that is, $|z| \le |x|$ and therefore it follows that $|y| \le |x| \Rightarrow y \in Z(x) \subseteq \bigcup_{x \in A} Z(x)$.)

Now if B is a solid subset of E such that $B \supseteq A$, then $B \supseteq \bigcup_{x \in A} Z(x)$.

(Indeed, if $z \in \bigcup_{x \in A} Z(x)$, then there exists $x \in A \subseteq B$ with $z \in Z(x)$, that is, $|z| \leq |x|$; it follows that $z \in B$, because B is a solid set.) □

Proposition 5.9. (see, for example, [AB1, p.322]) *For a (nonempty) subset A of E, the ideal generated by A can be described as*

$$I(A) = \left\{ x \in E \,\middle|\, \exists n \in \mathbb{N}^*, \exists x_1,...,x_n \in A, \exists \lambda_1 > 0,...,\lambda_n > 0 \text{ with } |x| \leq \sum_{i=1}^{n} \lambda_i |x_i| \right\}$$

Proof. It suffices to prove that the set on the right hand side is an ideal and the smallest one which contains the set A. □

Proposition 5.10. *For E a vector lattice and $x \in E$, the principal ideal generated by x is*

$$E_x = \left\{ y \in E \,\middle|\, \exists \lambda > 0 \text{ with } |y| \leq \lambda |x| \right\}.$$

Let us remark a special case, namely when $E_e = E$ for an element $e \in E_+$.
In other words for each $x \in E$ there exists $\lambda > 0$ such that $|x| \leq \lambda e$.
Recall that in this case we say that e is an *order unit*, or simply, a *unit* of E.

6. (o)-DENSE SUBSPACES

Classical definitions

Definition 6.1. Let E be a vector lattice and $G \subseteq E$ a vector sublattice. We say that G is an *(o)-dense subspace* of E, if for all $x \in E$, $x > 0$,

$$x = \sup \left\{ y \in G \,\middle|\, 0 \leq y \leq x \right\}.$$

As an example of an (o)-dense subspace of an Archimedean vector lattice E, we notice that any *ideal* G of E which moreover is *total*, that is, $G^\perp = \{0\}$, is an *(o)-dense subspace* - see [C1, p.96]. Recall that $G^\perp$, called the *orthogonal complement* of G, is the following set:

$$G^\perp = \left\{ x \in E \,\middle|\, x \perp v, \text{ for all } v \in G \right\}.$$

Algebraic description

By using Proposition 1.3 it follows:

Proposition 6.1. *A sublattice G of a vector lattice E is an (o)-dense subspace of E, if and only if for all $x \in E$, $x > 0$*

$$x + E_+ = \bigcap_{\substack{y \in G_+ \\ x \in y + E_+}} (y + E_+)$$

or, equivalently,

$$x + E_+ = \bigcap_{y \in G_+ \cap (x - E_+)} (y + E_+).$$

Proof. Let G be a sublattice of the vector lattice E. We know that G is also an (o)-dense subspace of E, that is, for all $x \in E$, $x > 0$ it follows that

$$x = \sup\{y \in G \mid y \in E_+ \cap (x - E_+)\}. \tag{25}$$

Let $a \in E_+$, hence $x + a \in x + E_+$.

From (25) it follows that $y + a \in x + a - E_+$ for all $y \in G$ with $y \in E_+ \cap (x - E_+) \Rightarrow$

$$x + a \in (y + a) + E_+ \subseteq (y + E_+) + E_+ = y + E_+$$

for all $y \in G$ with $y \in E_+ \cap (x - E_+)$.

Hence $x + a \in \bigcap_{\substack{y \in G_+ \\ x \in y + E_+}} (y + E_+)$.

Conversely if $z \in \bigcap_{\substack{y \in G_+ \\ x \in y + E_+}} (y + E_+)$ then, for all $y \in G_+$ such that $x \in y + E_+$ there exists $b \in E_+$ with $z = y + b$. Because G is an (o)-dense subspace of E, it follows that $z \in x + E_+$.

The converse implication: Let $x \in E$, $x > 0$. We have to prove that x is the supremum of the set $\{y \in G \mid 0 \le y \le x\}$. Obviously x is an upper bound of this set.

Now, let $z \in y + E_+$ for all $y \in G_+$ with $x \in y + E_+$. Because $x + E_+ = \bigcap_{\substack{y \in G_+ \\ x \in y + E_+}} (y + E_+)$ it follows that $z \in x + E_+$, that is, $z \ge x$.

Hence $z = \sup\{y \in G \mid 0 \le y \le x\}$. □

The following proposition can be easily proved with the previous result.

Proposition 6.2. *Let E be a vector lattice and $G \subseteq E$ a vector subspace. Then, G is an (o)-dense subspace of E if and only if, for all $x \in E$, $x > 0$, there exists $y \in E$, $y \ne 0$ with $y \in G_+ \cap (x - E_+)$.*

Proof. Suppose that G is an (o)-dense subspace of E. Then, obviously, for all $x \in E$, $x > 0$, there exists $y \in E$ with $y \in G_+ \cap (x - E_+)$.

Conversely, let $A = \{y \in G \mid 0 \le y \le x\}$. Obviously according to the hypothesis of this implication, A is a nonempty subset of G. It must to prove that $x = \sup A$, that is, $x + E_+ = \bigcap_{y \in A}(y + E_+)$. Obviously, for all $y \in A$, $x \in y + E_+$ $\Rightarrow x + E_+ \subseteq \bigcap_{y \in A}(y + E_+)$. For the converse inclusion, let $z \in E$ such that $z \in y + E_+$ for all $y \in A$. We have to prove that $z \in x + E_+$. We remark that putting $z \wedge x$ instead of z we also have $z \wedge x \in y + E_+$ for all $y \in A$ and at the same time, $z \wedge x \in x - E_+$. Therefore we will prove that if $y_0 \in y + E_+$ for all $y \in A$ and $y_0 \in x - E_+$, then $y_0 = x$. Suppose by the way of contradiction that $y_0 < x$. Then $u = x - y_0 > 0$ and, according to the hypothesis, there exists $y_1 \in E$, $y_1 \ne 0$ such that $y_1 \in G_+ \cap (u - E_+)$. It follows that $y_1 \in x - y_0 - E_+$ and then $y_0 + y_1 \in E_+ \cap (x - E_+)$. But $y_1 \in y_1 + y_0 - E_+$, hence $y_1 \in A$. Now let $y \in A$. It follows that $y \in y_0 - E_+$ and $y + y_1 \in E_+ \cap (y_0 + y_1 - E_+) \Rightarrow y + y_1 \in E_+ \cap (x - E_+)$. Hence, for all $y \in A$, it follows that $y + y_1 \in A$. In particular, taking $y = y_1$, it follows that $2y_1 \in A$. By induction we will obtain that $ny_1 \in A$, for all $n \in \mathbb{N}^*$. Therefore

$$ny_1 \in x - E_+,$$

for all $n \in \mathbb{N}^*$ and hence $y_1 \le 0$ (E being Archimedean). But then $y_1 \in -E_+ \cap E_+ = \{0\}$ and this is in contradiction with $y_1 \ne 0$. Hence, it follows that $y_0 = x$, that is, $x = \sup A$. □

7. VARIOUS TYPES OF LINEAR OPERATORS COMMUTING WITH LATTICE OPERATIONS

Classical definitions

Definition 7.1. Let E and F be two vector lattices, $G \subseteq E$ a sublattice and $T : E \to F$ a linear operator. We say that T is a *G-Riesz homomorphism* (in short, a *Riesz homomorphism*) if and only if

$$T(u \vee v) = T(u) \vee T(v) \text{ for all } u, v \in G. \tag{26}$$

(For a short paper concerning Riesz homomorphisms see, for example, [D4].)

Definition 7.2. Let E and F be two vector lattices, $G \subseteq E$ a lattice-subspace and $T : E \to F$ a linear operator. We say that T is a *G-lattice operator* (in short, a *restricted-lattice operator*) if and only if

$$T(u \vee_G v) = T(u) \vee T(v) \text{ for all } u, v \in G. \tag{27}$$

(This notion was introduced in [D3] and used in [D5].)

Definition 7.3. Let E and F be two vector lattices, $G \subseteq E$ a lattice-subspace and $T : E \to F$ a linear operator, such that $T(G)$ is a lattice-subspace of F. We say that T is a *G-quasi lattice operator* if and only if

$$T(u \vee_G v) = T(u) \vee_{T(G)} T(v) \text{ for all } u, v \in G. \tag{28}$$

According to the Remark before Proposition 4.1, any sublattice $G \subseteq E$ is a lattice-subspace and in this case $u \vee_G v = u \vee v$ for all $u, v \in G$.

We can prove the following result.
Proposition 7.1. *Let E and F be two vector lattices and $G \subseteq E$ a sublattice. Let also $T : E \to F$ be a G-Riesz homomorphism. Then:*
a) $T(G)$ is a sublattice in F ;
b) T is a G-lattice operator, and a G-quasi lattice operator, too.

Proof.
a). Let $T(u), T(v) \in T(G)$. Since G is a sublattice and $u, v \in G$, we have:

39

$$T(u) \vee T(v) \overset{\underset{\text{homomorphism}}{T \text{ is Riesz}}}{=} T(u \vee v) \in T(G).$$

Therefore, $T(G)$ is a sublattice.

b). Now G being sublattice is a G lattice-subspace too, and then:

$$T(u \vee_G v) = T(u \vee v) \overset{\underset{\text{homomorphism}}{T \text{ is Riesz}}}{=} T(u) \vee T(v) \tag{29}$$

for all $u, v \in G$. So, T is a G-lattice operator.

We also have:

$$T(u) \vee_{T(G)} T(v) = T(u) \vee T(v) \tag{30}$$

(because G being a sublattice, from a) it follows that $T(G)$ is a sublattice, too; therefore $T(G)$ is a lattice-subspace). Now, we have:

$$T(u \vee_G v) = T(u \vee v) \overset{\underset{\text{homomorphism}}{T \text{ is Riesz}}}{=} T(u) \vee T(v) \overset{(30)}{=} T(u) \vee_{T(G)} T(v)$$

for all $u, v \in G$. So, T is a G-quasi lattice operator. $\qquad\qquad \square$

Algebraic descriptions

Let E and F be two vector lattices and $G \subseteq E$ a sublattice. Let also $T : E \to F$ be a linear positive operator.

Proposition 7.2. *The operator T is a G-Riesz homomorphism if and only if*
$$T(u \vee v + E_+) = T(u) \vee T(v) + T(E_+) \text{ for all } u, v \in G. \tag{31}$$

Proof.

"$\Rightarrow$" "$\subseteq$" Let $z \in E_+$. Therefore $T(u \vee v + z) \in T(u \vee v + E_+)$ and

$$T(u \vee v + z) \overset{\underset{\text{homomorphism}}{T \text{ is Riesz}}}{=} T(u) \vee T(v) + T(z) \in T(u) \vee T(v) + T(E_+).$$

"$\supseteq$" Conversely, is obvious.

"$\Leftarrow$" Take $0 \in E_+ \Rightarrow 0 = T(0) \in T(E_+) \Rightarrow$

$$\Rightarrow T(u) \vee T(v) \in T(u) \vee T(v) + T(E_+) \overset{(31)}{=} T(u \vee v + E_+) \Rightarrow$$
$$\Rightarrow T(u) \vee T(v) \in T(u \vee v + E_+) \Rightarrow$$

40

$$\Rightarrow \exists\, z \geq 0 \text{ such that } T(u) \vee T(v) = T(u \vee v + z)$$

$$\overset{T \text{ linear}}{\Rightarrow}\ T(u) \vee T(v) = T(u \vee v) + \underbrace{T(z)}_{\substack{\geq 0 \\ (T \text{ positive})}}$$

$$\Rightarrow T(u) \vee T(v) \geq T(u \vee v) \tag{32}$$

Now, because $u,v \leq u \vee v \overset{T \text{ positive}}{\Rightarrow} T(u), T(v) \leq T(u \vee v)$ and hence

$$T(u) \vee T(v) \leq T(u \vee v) \tag{33}$$

From (32) and (33) it follows that

$$T(u \vee v) = T(u) \vee T(v). \qquad \square$$

By using Proposition 7.2, the following lemma and the identity (2) written for u and v $(u,v \in E)$ instead of x and y, respectively, that is,

$$u \vee v + E_+ = (u + E_+) \cap (v + E_+) \tag{2}$$

we obtain for any *surjective E-Riesz homomorphism* $T : E \to F$:

$$T\big((u + E_+) \cap (v + E_+)\big) = \big(T(u) + F_+\big) \cap \big(T(v) + F_+\big) \tag{34}$$

Lemma 7.3. *If T is a surjective E-Riesz homomorphism, then*

$$T(E_+) = F_+ \big(= T(E)_+\big).$$

Proof of Lemma.

" $\subseteq$ " Let T be an E-Riesz homomorphism $\Rightarrow T$ positive $\Rightarrow T(E_+) \subseteq F_+$

" $\supseteq$ " Conversely, let $y \in F_+ \overset{T \text{ surjective}}{\Rightarrow} \exists\, x \in E$ such that $y_+ = y = T(x)$.
Remark that we can choose $x \geq 0$. Indeed, we have

$$T(x) = y = y_+ = \big(T(x)\big)_+ \overset{\substack{T \text{ is a } E-Riesz \\ homomorphism}}{=} T(x_+)$$

Hence, instead of x, we can take x_+ and therefore $y \in T(E_+)$. $\qquad \square$

Proposition 7.4. *If the operator $T : E \to F$ is a surjective E-Riesz homomorphism, then the identity (34) is valid for all $u,v \in E$.*

Proof. From the identities (2), (31) and again (2) written for $T(u)$ and $T(v)$

instead of u and v, respectively we have:

$$T\big((u+E_+)\cap(v+E_+)\big)\overset{(2)}{=}T(u\vee v+E_+)\overset{(31)}{=}T(u)\vee T(v)+T(E_+)=$$

$$\overset{Lemma\ 7.3}{=}\ T(u)\vee T(v)+F_+\overset{(2)}{=}\big(T(u)+F_+\big)\cap\big(T(v)+F_+\big).\qquad\square$$

Similarly with Proposition 7.3 and Proposition 7.4, respectively, we have the following two results:

Proposition 7.5. *Let E and F be two vector lattices, $G\subseteq E$ a lattice-subspace and $T:E\to F$ a linear positive operator. T is a G-lattice operator if and only if*

$$T(u\vee_G v+G_+)=T(u)\vee T(v)+T(G_+).$$

Proposition 7.6. *Let E and F be two vector lattices, G a lattice-subspace and $T:E\to F$ a G-lattice operator such that $S=T|_G$ is surjective. Then:*

$$T\big((u+G_+)\cap(v+G_+)\big)=\big(T(u)+F_+\big)\cap\big(T(v)+F_+\big).$$

8. AN EXAMPLE

Consider $E=C[0,1]$, and its vector subspace $G=\big\{f\in E\big|f\ \text{affine on}\ [0,1]\big\}$.

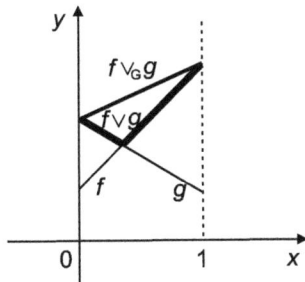

It is known that G is a lattice-subspace of E, but it is not a sublattice of E.

We have $f\vee g\leq f\vee_G g$, $\forall f,g\in G$.

42

Define $T : E \rightarrow E$ by $T(f) = f$, for all $f \in E$ (the *identity operator*).

It follows that:
1) T is an E-Riesz homomorphism, that is, for all $f, g \in E$:
$$T(f \vee g) = f \vee g = T(f) \vee T(g).$$
2) T is a G-quasi lattice operator, that is, for all $f, g \in G$:
$$T(f \vee_G g) = f \vee_G g \overset{G=T(G)}{=} T(f) \vee_{T(G)} T(g)$$

But the operator T is not necessarily a G-lattice operator. Indeed, it is possible that there exist $f, g \in G$ such that
$$T(f \vee_G g) \neq T(f) \vee T(g)$$
or, equivalently,
$$f \vee_G g \neq f \vee g.$$

9. SOMETHING ABOUT RIESZ HOMOMORPHISMS

In this section we have in view the Riesz homomorphisms $T : E \rightarrow F$ with E and F two vector lattices, that is, the E-Riesz homomorphisms.

Note that the equality $x_1 + x_2 = x_1 \vee x_2 + x_1 \wedge x_2$, valid for all x_1, x_2 in E, shows us that if T preserves "$\vee$"(or "$\wedge$"), then it preserves "$\wedge$" (or "$\vee$"), too.

A short overview of these operators is made in [D4].

The following result gives other descriptions for a Riesz homomorphism $T : E \rightarrow F$.

Proposition 9.1. ([D4, Proposition 1], [J, 2.6.6], [AB2, Theorem 1.17], [S, p.59]) *Let E and F be two vector lattices and $T : E \rightarrow F$ a linear operator. Then, the following are equivalent:*
i) *T preserves the vector lattices operations (sup and inf), that is:*

$$T(x_1 \vee x_2) = T(x_1) \vee T(x_2), \text{ and}$$
$$T(x_1 \wedge x_2) = T(x_1) \wedge T(x_2) \text{ for all } x_1, x_2 \text{ in } E ;$$

ii) $T(x^+) \wedge T(x^-) = 0$ for each $x \in E$;

iii) If $x_1 \wedge x_2 = 0$ in E, then $T(x_1) \wedge T(x_2) = 0$ in F (that is T is disjunctive);

iv) If $x_1 \wedge x_2 = 0$ in E, then $T(x_1 \vee x_2) = T(x_1) \vee T(x_2)$;

v) $T(x^+) = (T(x))^+$ for any $x \in E$;

vi) $T(|x|) = |T(x)|$ for any $x \in E$.

Proof.

i) $\Rightarrow$ ii). $T(x^+) \wedge T(x^-) \overset{\text{i)}}{=} T(x^+ \wedge x^-) = T(0) = 0$ (because T is linear).

ii) $\Rightarrow$ iii). If $x_1 \wedge x_2 = 0$ in E, then:
$$(x_1 - x_2)^+ = x_1 \vee x_2 - x_2 = x_1 - x_1 \wedge x_2 = x_1$$
and
$$(x_1 - x_2)^- = x_1 \vee x_2 - x_1 = x_2 - x_1 \wedge x_2 = x_2 .$$

So,
$$T(x_1) \wedge T(x_2) = T\left((x_1 - x_2)^+\right) \wedge T\left((x_1 - x_2)^-\right) \overset{\text{ii)}}{=} 0$$

iii) $\Rightarrow$ iv). $x_1 + x_2 = x_1 \vee x_2 + x_1 \wedge x_2 = x_1 \vee x_2$ because $x_1 \wedge x_2 = 0$.

Then:
$$T(x_1 \vee x_2) = T(x_1) + T(x_2) \overset{\text{iii)}}{=} T(x_1) + T(x_2) - T(x_1) \wedge T(x_2) =$$
$$= T(x_1) \vee T(x_2).$$

iv) $\Rightarrow$ iii). It is immediate.

iii) $\Rightarrow$ i). Because
$$(x_1 - x_1 \wedge x_2) \wedge (x_2 - x_1 \wedge x_2) = x_1 \wedge x_2 - x_1 \wedge x_2 = 0 .$$

Then, by using iii), it follows that
$$T(x_1 - x_1 \wedge x_2) \wedge T(x_2 - x_1 \wedge x_2) = 0 ,$$

hence:
$$\left(T(x_1) - T(x_1 \wedge x_2)\right) \wedge \left(T(x_2) - T(x_1 \wedge x_2)\right) = 0$$
$$\Leftrightarrow T(x_1) \wedge T(x_2) - T(x_1 \wedge x_2) = 0 \Rightarrow \text{ i)}.$$

i) $\Rightarrow$ v). It is easily checked.

v) $\Rightarrow$ vi). Obviously, since $|x| = 2x^+ - x$, for any $x \in E$.

vi) $\Rightarrow$ v). Clearly, because $x^+ = \dfrac{1}{2}(|x| + x)$.

v) $\Rightarrow$ i). One uses the equality

$$x_1 \vee x_2 = (x_1 - x_2)^+ + x_2,$$

valid for any $x_1, x_2 \in E$. $\qquad\qquad\qquad\qquad\qquad\qquad\qquad$ □

Remark that in [C2, p.249] it is shown that an additive operator $T : E \to F$ satisfies i) if and only if T is *disjunctive* and *positive*. (Recall that T is called *positive* if $x \geq 0$ in E implies $T(x) \geq 0$ in F.)

We will denote by $H(E, F)$ the set of all Riesz homomorphisms $T : E \to F$.

The next four results (that is Proposition 9.2, 9.3, 9.4 and 9.5) give some *properties* for $T \in H(E, F)$.

Recall that an *ideal* in a vector lattice E is a vector subspace $G \subseteq E$ such that if $|y| \leq |x|$ and $x \in G$, then $y \in G$, too.

In the following proposition, we will denote by $\operatorname{Ker} T$ the kernel of the linear operator T (that is $\operatorname{Ker} T = T^{-1}(\{0\})$) and by N_T the set $\{x \in E \,|\, |T|(|x|) = 0\}$.

Proposition 9.2. (The properties of a Riesz homomorphism - see [D4, Proposition 2], [AB2], [J], [LZ], [S])
If $T \in H(E, F)$, then

1) $T\left(\displaystyle\bigvee_{i=1}^{n} x_i\right) = \displaystyle\bigvee_{i=1}^{n} T(x_i)$, *and*

$T\left(\displaystyle\bigwedge_{i=1}^{n} x_i\right) = \displaystyle\bigwedge_{i=1}^{n} T(x_i)$ *for all $n \in \mathbb{N}^*$ and $x_1, ..., x_n \in E$;*

2) *T is an increasing operator and, particularly, a positive operator;*
3) *If there exists $T^{-1} : E \to F$, then T^{-1} is also a Riesz homomorphisms and, actually, a positive operator;*

45

4) $\operatorname{Ker} T = N_T$ and $\operatorname{Ker} T$ is an ideal of E.

5) If $0 \le S \le T$ $(S : E \to F$ linear$)$, then $S \in H(E,F)$.

Proof.

2). If $x_1 \le x_2$ then $x_2 = x_1 \vee x_2$ and so $T(x_2) = T(x_1 \vee x_2) = T(x_1) \vee T(x_2)$, that is $T(x_1) \le T(x_2)$.

3). If $y_1, y_2 \in F$ such that $y_1 = T(x_1)$ and $y_2 = T(x_2)$. Then:
$$T^{-1}(y_1 \vee y_2) = T^{-1}\left(T(x_1) \vee T(x_2)\right) = T^{-1}\left(T(x_1 \vee x_2)\right) =$$
$$= x_1 \vee x_2 = T^{-1}(y_1) \vee T^{-1}(y_2).$$

4). $x \in \operatorname{Ker} T \Leftrightarrow T(x) = 0 \Leftrightarrow 0 = |T(x)| = T(|x|) = |T|(|x|) \Leftrightarrow x \in N_T$.

5). If $x_1 \wedge x_2 = 0$ in E, then $0 \le S(x_1 \wedge x_2) \le T(x_1 \wedge x_2) = 0$ in $F \Rightarrow$ $S(x_1 \wedge x_2) = 0$. $\qquad\square$

Remark. It is well known (see, for example, [D4, Counterexample 1]) that, generally, a Riesz homomorphism does not necessarily preserve infima or suprema of countable sets (though it is elementary that a *Riesz homomorphism* between two vector lattice preserves all finite suprema and infima).

Based on the above observation we can define the following notion.

Definition 9.1. A Riesz homomorphism $T : E \to F$ is called a (σ)-*Riesz homomorphism* (or an (o)-*continuous operator* - see [C1]) if it preserves suprema of countable sets:
$$T\left(\bigvee_{n \in \mathbb{N}} x_n\right) = \bigvee_{n \in \mathbb{N}} T(x_n) \text{ for all } (x_n)_{n \in \mathbb{N}} \subset E.$$

Definition 9.2. A Riesz homomorphism $T : E \to F$ is called a *normal Riesz homomorphism* (or an ω-*continuous operator* - see [C1]) if it preserves all suprema:
$$T\left(\bigvee_{\delta \in \Delta} x_\delta\right) = \bigvee_{\delta \in \Delta} T(x_\delta) \text{ for all } (x_\delta)_{\delta \in \Delta} \subset E.$$

The following result refer to an *onto Riesz homomorphism* $T : E \to F$ (that is, T is Riesz homomorphism such that $T(E) = F$).

Proposition 9.3. (the properties of an onto Riesz homomorphism) - see [D4, Proposition 3], [AB2] and [LZ].

If $T : E \to F$ is a linear operator between two vector lattices, such that $T(E_+) = F_+$, then:

a) *The following are equivalent:*

　　i) T *is a Riesz homomorphism;*

　　ii) $T^{-1}(H)$ *is a solid subset of E if H is a solid subset of F, and hence for all $|u| \le |v|$ with $u \in E$ and $v \in T^{-1}(H)$ it follows that $u \in T^{-1}(H)$.*

　　iii) $\operatorname{Ker} T$ *is an ideal of E.*

b) *If T is a Riesz homomorphism, then it carries solid subsets of E onto solid subsets of F.*

c) *If T is a Riesz homomorphism, then it carries the ideals of E onto the ideals of F.*

d) *If T is a Riesz homomorphism, then*

$$T(G_1 \cap G_2) = T(G_1) \cap T(G_2)$$

for any two ideals G_1, G_2 of E.

Remarks.

1) Obviously, if $T(E_+) = F_+$, then T is onto and for T a Riesz homomorphism, the converse is also true.

2) Also, related to "i) $\Leftrightarrow$ iii)", we can prove that, moreover, if $T : E \to F$ is an onto linear operator, then "i) $\Leftrightarrow$ iii')", where:

　　iii') $\operatorname{Ker} T$ is an ideal of E and $T(E_+) = F_+$ (see, for example, [J, p.73]).

Proof.

a). "i) $\Leftrightarrow$ ii)". We use Proposition 9.1, "i) $\Leftrightarrow$ vi)". (Firstly we remark that, because T is *onto*, then $T^{-1}(H) \ne \varnothing$ for all $H \ne \varnothing$ in F.)

Indeed, if $x_1 \in T^{-1}(H)$ and $x_2 \in E$ with $|x_2| \le |x_1|$, then

$$|T(x_2)| = T(|x_2|) \le T(|x_1|) \in H$$

and so $T(x_2) \in H$, that is, $x_2 \in T^{-1}(H)$.

"ii) $\Leftrightarrow$ iii)". It is immediate.

"iii) $\Leftrightarrow$ i)". Denote $G = \operatorname{Ker} T$ and let $\varphi : E \to E/G$ be the canonical map. (It is known that under the finest ordering of E/G for which φ is positive, E/G is a vector lattice and φ is a Riesz homomorphism onto E/G - see, for example, [S, p.59]). Let $S : E/G \to F$ be defined by $S(\varphi(x)) = T(x)$ for all $x \in E$. It follows that S is an isomorphism in the setting of all vector lattices. Since φ preserves the lattice operations, T also preserves these operations.

b). If M is a solid subset of E, $0 \le v \in M$ and $0 \le z \le T(v)(\in T(M))$, then there exists $u \in E$ such that $z = T(u_+)$. But, the element $w = v \wedge u_+$ belongs to M and $T(w) = T(v) \wedge T(u_+) = T(v) \wedge z = z$. Hence, $z \in T(M)$ and thus $T(M)$ is a solid subset of F.

c). It is immediate.

d). We have to show only the inclusion " $\supseteq$ ". If $y \in T(G_1) \cap T(G_2)$, then according to b), $|y| \in T(G_1) \cap T(G_2)$ and thus there exist $x_1 \in G_1 \cap E_+$, $x_2 \in G_2 \cap E_+$ such that $T(x_1) = |y| = T(x_2)$. Now, if $u = x_1 \wedge x_2$, then $0 \le u \in G_1 \cap G_2$ and $T(u) = |y|$. But, according b), $T(G_1 \cap G_2)$ is a solid subset of F and hence $y \in T(G_1 \cap G_2)$. $\qquad\qquad\square$

The following result characterizes the onto Riesz homomorphisms in the case $F = \mathbb{R}$ (that is the case of the onto Riesz functionals).

Proposition 9.4. ([D4, Proposition 4], [S, p.74]) *Let E be a vector lattice and $f : E \to \mathbb{R}$ a linear functional such that $f \ne 0$. Then, the following are equivalent:*
i) f is an onto Riesz homomorphism;
ii) $f(x^+) \wedge f(x^-) = 0$ for all $x \in E$;
iii) $f \ge 0$ and $\mathrm{Ker}\, f$ is a maximal ideal of E;
iv) $f \ge 0$ and the (principal) ideal generated by f in $H(E,\mathbb{R})$, that is, the subspace

$$E'_f = \left\{ g \in H(E,\mathbb{R}) \mid \exists\, \lambda > 0, |g| \le \lambda |f| \right\}$$

is a minimal subspace (and, therefore, $\dim E'_f = 1$).

Proof.

"i)$\Leftrightarrow$ ii)$\Leftrightarrow$ iii)". These equivalences are immediate consequences of Proposition 9.1 and Remark 2 after Proposition 9.3.

"i)$\Rightarrow$iv)". We will show that $\{f\}$ is a base in E'_f. Some authors - for example W.A.J. Luxemburg and A.C Zaanen - see [LZ], call such a functional f as a *discrete* element of E'. Other authors, for example, G. Jameson -see [J], call such a functional an *extreme element* of E' (more general, an *extreme functional* in E' is an element $0 \leq f \in E'$ such that for any $g \in [0,f]$ there exists $\lambda \geq 0$ such that $g = \lambda f$). To prove that $\{f\}$ is a base in E'_f, we choose $g \in E'_f$. It follows that there exists $\lambda > 0$ such that $|g| = \lambda f \Rightarrow \operatorname{Ker} f \subseteq \operatorname{Ker} g$. But $\operatorname{Ker} f$ is a maximal ideal in E. It follows that either $\operatorname{Ker} g = E$, that is, $g = 0$, or $\operatorname{Ker} g = \operatorname{Ker} f \Rightarrow$ there exists $\mu \in \mathbb{R}$ such that $g = \mu f$. Therefore, $\{f\}$ is a base in E'_f. It follows that $\dim E'_f = 1 \Rightarrow E'_f$ is a minimal ideal of E'.

"iv)$\Rightarrow$ ii)". If $x \in E$ and $f \geq 0$, then $f(x^+) \geq 0$ and $f(x^-) \geq 0$. Let us suppose that $f(x^-) > 0$. Let P be the convex cone $\bigcup_{n \geq 1} n[0, x^-]$ and $g : E_+ \to \mathbb{R}_+$ defined by

$$g(y) = \sup\{f(z) \mid z \in [0, y] \cap P\}$$

for any $y \in E_+$. But E, being a vector lattice, has the *Riesz Decomposition Property*, that is,

$$[0, y_1 + y_2] \cap P = [0, y_1] \cap P + [0, y_2] \cap P.$$

It follows from this that g is additive on E_+. Obviously g is positively homogeneous, too. Therefore g can be extended to a linear functional $g : E \to \mathbb{R}$ such that $0 \leq g \leq f$. Then $g \in E'_f$ and by using "iv)" we can choose $\lambda \geq 0$ such that $g = \lambda f$. Because $g(x^-) = f(x^-) > 0$ it follows that $\lambda = 1$ and hence, $g = f$. But $g(x^+) = 0$ and thus $f(x^+) = 0$; therefore $f(x^+) \wedge f(x^-) = 0$. $\qquad\square$

Concerning the above characterization of the onto Riesz functionals (see Proposition 9.4 i)$\Leftrightarrow$ iv)) we recall the following:

Definition 9.3. (see [S, p.67]) An element $x \neq 0$ of a vector lattice E is called an *atom* if the (principal) ideal generated by x, that is E'_x, is totally ordered. This concept is mainly useful for Archimedean vector lattices.

Lemma 9.5. *If E is Archimedean, then the following are equivalent:*
    i) x is an atom of E ;
    ii) E'_f is minimal;

    iii) The orthogonal complement $\left(E'_f\right)^{\perp}$ of E'_f is maximal.

Remark. By using Lemma 9.5 and Proposition 9.4 (i) $\Leftrightarrow$ iv)), we can now show that if f is an onto Riesz functional, then $f \geq 0$, f being an atom of E', and conversely too. Also we can characterize the Riesz functionals $f : E \to \mathbb{R}$ among the nontrivial positive linear functionals $f : E \to \mathbb{R}$, which are *order bounded*. (Recall that f is *order bounded* if f carries order intervals of E to bounded subsets of $\mathbb{R}$.)

We denote by $E^{\#}$ the collection of all order bounded functionals on E. According to [C1, p.167], for example, $E^{\#}$ coincides with the regular dual E^r of E. Recall that $f : E \to \mathbb{R}$ is a *regular functional* if f can be written as a difference of two positive linear functionals (see, for example [AT, p.31]).

Proposition 9.6. (see [D4, Proposition 6], [AB2, Theorem 3.13]) *For E a vector lattice and $0 < f \in E^{\#}$ the following statements are equivalent:*
    i) f is a Riesz homomorphism (from E into $\mathbb{R}$);
    ii) f is a discrete element of E (that is $E'_f = \{\lambda f \mid \lambda \in \mathbb{R}\}$).
Proof.
"i) $\Rightarrow$ ii)". Let $g \in E^{\#}$ such that $0 \leq g \leq f$. Then:
$$\text{Ker } f \subseteq \text{Ker } g \Rightarrow \exists\, \lambda \in \mathbb{R}, \text{ with } f = \lambda g.$$
"ii) $\Rightarrow$ i)". Firstly, because $|g| \leq f$ implies $g = \lambda f$ for some $\lambda \leq 1$, we have:
$$|g(x)| = |\lambda f(x)| = |\lambda||f(x)| \leq |f(x)|$$
for all $x \in E$, and hence
$$f(|x|) = \sup\{|g(x)| \mid g \in E^{\#}, |g| \leq f\} \leq |f(x)|$$

for all $x \in E$. Because $|f(x)| \le f(|x|)$ it follows that $f(|x|) = |f(x)|$ for all $x \in E$, that is, f is a Riesz functional. □

In what follows we recall some basic properties of Riesz isomorphisms.

Definition 9.4. A Riesz homomorphism $T : E \to F$ is called a *Riesz isomorphism* if it is bijective.

From Proposition 9.2 ("3)"), it follows that if $T : E \to F$ is a Riesz homomorphism, then the inverse operator $T^{-1} : F \to E$ is also a Riesz homomorphism.

Definition 9.5. The vector lattices E and F are called *Riesz isomorphic* if there exists a Riesz isomorphism $T : E \to F$.

Remark. Obviously, any Riesz isomorphism $T : E \to F$ is a bijective positive linear operator, but the converse is not true.

Counterexample. ([D4]) Let $E = \mathbb{R}^2$, endowed with the canonical ordering:
$$(x_1, y_1) \le_E (x_2, y_2) \text{ if } x_1 \le x_2 \text{ and } y_1 \le y_2.$$
Let also $F = \mathbb{R}^2$, endowed with the lexicographic ordering, that is
$$(x_1, y_1) \le_F (x_2, y_2) \text{ if } x_1 < y_1 \text{ or } x_1 = y_1 \text{ and } x_2 \le y_2.$$
The identity operator $T : E \to F$ is obviously a bijective positive linear operator, but it is not a Riesz homomorphism because
$$T^{-1}(F_+) = F_+ \not\subseteq E_+.$$
Therefore T^{-1} is not a positive operator (we apply again Proposition 9.2 ("3)"). Moreover, the spaces E and F are not Riesz isomorphic. (Indeed E is Archimedean, but F is not.) Otherwise any one to one linear operator on E onto F transforms the halfplane $\{(x_1, x_2) \mid x_1 > 0\}$ into a halfplane, so there must be positive elements of E that are transformed into non-positive elements of F.) The conclusion of this counterexample is that we can seek a characterization of the Riesz isomorphism among the one to one and onto positive linear operators $T : E \to F$. Indeed, the following result is true.

Proposition 9.7. (see [LZ, Thm. 18.5]) *The one to one positive linear operator*

T of E onto F is a Riesz homomorphism if and only if T^{-1} is also positive.

Proof. Remark that, according Proposition 9.2 ("3)"), T^{-1} is positive if T is a bijective Riesz homomorphism. Conversely, assume that T^{-1} is also positive. We have to show that

$$T(x_1 \vee x_2) = T(x_1) \vee T(x_2) \text{ for all } x_1, x_2 \in E.$$

The inequality "$\geq$" is immediate, because $T \geq 0$.
The *converse* inequality: from

$$T(x_1 \vee x_2) \geq T(x_1), T(x_2)$$

it follows that

$$T^{-1}(T(x_1) \vee (x_2)) \geq x_1 \vee x_2 \text{ (since } T^{-1} \text{ is positive)}$$

and so

$$T(x_1) \vee (x_2) \geq T(x_1 \vee x_2) \text{ (since } T \geq 0 \text{)}. \qquad \square$$

In the remainder part of this section, we recall a property of a Riesz isomorphism pointed out in the Remark after Proposition 9.2: such an operator preserves all suprema and infima. Indeed, if $T : E \to F$ is a Riesz isomorphism and $G \subseteq E$ has the supremum v, then for all $x \in G \Rightarrow x \leq v \Rightarrow T(x) \leq T(v)$.

Also, if $T(w) \geq T(x)$ for all $x \in G$, then

$$T^{-1}(T(w)) \geq T^{-1}(T(x)),$$

that is $w \geq x$; hence $w \geq v$, or, equivalently,

$$T(w) \geq T(v).$$

So, it follows that $T(\sup G) = \sup T(G)$.

10. ON RESTRICTED-LATTICE OPERATORS

The following result is in the line of [D4, Proposition 1, p.47] (concerning Riesz homomorphisms - see also Proposition 9.1 in this paper).

Proposition 10.1. *Let E and F be two vector lattices, $G \subseteq E$ a lattice-subspace and $T : E \to F$ a linear operator. The following are equivalent:*

(i) *T preserves the lattice operations considered in G and F, that is, T is a G-lattice operator (or, equivalently, a restricted-lattice operator);*

(ii) $T\left(u_G^+\right)\wedge T\left(u_G^-\right)=0,\ \forall u\in G$;

(iii) If $u_1\wedge_G u_2=0\ \left(u_1,u_2\in G\right)\Rightarrow T\left(u_1\right)\wedge T\left(u_2\right)=0$ in F (we say that T is G-disjunctive);

(iv) If $u_1\wedge_G u_2=0\Rightarrow T\left(u_1\vee_G u_2\right)=T\left(u_1\right)\vee T\left(u_2\right)$;

(v) $T\left(u_G^+\right)=\left(T\left(u\right)\right)^+$ for all $u\in G$;

(vi) $T\left(\left.|u|\right._G\right)=\left|T\left(u\right)\right|$ for all $u\in G$.

Proof.

$(i)\Rightarrow(ii)$. $T\left(u_G^+\right)\wedge T\left(u_G^-\right)\overset{(i)}{=}T\left(u_G^+\wedge_G u_G^-\right)=T\left(0\right)=0$.

$(ii)\Rightarrow(iii)$. If $u_1\wedge_G u_2=0$ in G , then

$$\left(u_1-u_2\right)_G^+=u_1\vee_G u_2-u_2=u_1-u_1\wedge_G u_2=u_1\text{, and}$$

$$\left(u_1-u_2\right)_G^-=u_1\vee_G u_2-u_1=u_2-u_1\wedge_G u_2=u_2\text{.}$$

So:

$$T\left(u_1\right)\wedge T\left(u_2\right)=T\left(\left(u_1-u_2\right)_G^+\wedge\left(u_1-u_2\right)_G^-\right)\overset{(ii)}{=}0.$$

$(iii)\Rightarrow(iv)$. Since $u_1\wedge_G u_2=0$, then $u_1+u_2=u_1\vee_G u_2+u_1\wedge_G u_2=u_1\vee_G u_2$. Therefore

$$T\left(u_1\vee_G u_2\right)=T\left(u_1\right)+T\left(u_2\right)\overset{(iii)}{=}T\left(u_1\right)+T\left(u_2\right)-T\left(u_1\right)\wedge T\left(u_2\right)=T\left(u_1\right)\vee T\left(u_2\right).$$

$(iv)\Rightarrow(iii)$. It is immediate.

$(iii)\Rightarrow(i)$. Because $\left(u_1-u_1\wedge_G u_2\right)\wedge\left(u_2-u_1\wedge_G u_2\right)=u_1\wedge_G u_2-u_1\wedge_G u_2=0$, by using iii) it follows that

$$T\left(u_1-u_1\wedge_G u_2\right)\wedge T\left(u_2-u_1\wedge_G u_2\right)=0\ .$$

Hence

$$0=\left(T\left(u_1\right)-T\left(u_1\wedge_G u_2\right)\right)\wedge\left(T\left(u_2\right)-T\left(u_1\wedge_G u_2\right)\right)=$$

$$=T\left(u_1\right)\wedge T\left(u_2\right)-T\left(u_1\wedge_G u_2\right)\Rightarrow(i).$$

$(i)\Rightarrow(v)$. It easily checked.

$(v)\Rightarrow(vi)$. Obviously, since $\left.|u|\right._G=2u_G^+-u$ for any $u\in G$.

(vi) $\Rightarrow$ (v). Clearly, because $u_G^+ = \dfrac{1}{2}\left(|u|_G + u\right)$ for any $u \in G$.

(vi) $\Rightarrow$ (i). Use the equality $u_1 \vee_G u_2 = \left(u_1 - u_2\right)_G^- + u_2$ for any $u_1, u_2 \in G$. $\quad\square$

The following result is in the line of Proposition 9.2.

Proposition 10.2. *(Properties of a G-lattice operator).*
Let E and F be two vector lattices, $G \subseteq E$ a lattice-subspace and $T : E \to F$ a G-lattice operator. Then:

1) $T\left(\overset{n}{\underset{i=1}{\vee_G}} v_i\right) = \overset{n}{\underset{i=1}{\vee}} T\left(v_i\right)$, *and*

$T\left(\overset{n}{\underset{i=1}{\wedge_G}} v_i\right) = \overset{n}{\underset{i=1}{\wedge}} T\left(v_i\right)$ *for all $n \in \mathbb{N}^*$ and $v_1, v_2, ..., v_n \in G$.*

2) *T is monotone on G, and, in particular, T is positive on G.*

3) *If there exists $\left(T\restriction_G\right)^{-1} : T\left(G\right) \to G$, then $\left(T\restriction_G\right)^{-1}$ is a $T\left(G\right)$-lattice operator and, in fact, a positive operator.*

4) *If $0 \leq S \leq T$ on G, with $T, S : E \to F$ and T is a G-lattice operator, then S is a G-lattice operator, too.*

Proof.
2) If $u_1 \leq u_2$ in G, then $u_2 = u_1 \vee_G u_2$ and so

$$T\left(u_2\right) = T\left(u_1 \vee_G u_2\right) \overset{\substack{T\ is\ a\\ G-lattice\\ operator}}{=} T\left(u_1\right) \vee T\left(u_2\right) \Rightarrow T\left(u_1\right) \leq T\left(u_2\right).$$

3) Let $y_1, y_2 \in T\left(G\right) \Rightarrow$ there exist two unique elements $u_1, u_2 \in G$ such that $y_1 = T\left(u_1\right)$, $y_2 = T\left(u_2\right)$. Then:

$$T^{-1}\left(y_1 \vee y_2\right) = T^{-1}\left(T\left(u_1\right) \vee T\left(u_2\right)\right) \overset{\substack{T\ is\ a\\ G-lattice\\ operator}}{=} T^{-1}\left(T\left(u_1 \vee_G u_2\right)\right) =$$

$$= u_1 \vee_G u_2 = T^{-1}\left(y_1\right) \vee_G T^{-1}\left(y_2\right).$$

4) If $u_1 \wedge_G u_2 = 0$ in $G \Rightarrow 0 \leq S\left(u_1 \wedge_G u_2\right) \leq T\left(u_1 \wedge_G u_2\right) = 0$ (because T is a G-lattice operator) $\Rightarrow S\left(u_1 \wedge_G u_2\right) = 0$. $\quad\square$

Remark. The following example shows again what is *difference* between the *lattice operations calculated* in a *vector lattice* and, respectively, in a *lattice-subspace*. Also, this example proves that a Riesz homomorphism does not necessarily preserve infima or suprema of countable sets.

Take $E = C([0,1])$, endowed with the pointwise algebraic and order structures, $F = \mathbb{R}$, and $G = \{at + b \mid a,b \in \mathbb{R}, t \in [0,1]\}$, the space of all affine functions on $[0,1]$. For any $t \in [0,1]$, define $T_t : E \to F$ by $T_t(x) = x(t)$ for all $x \in E$. Obviously, for all $t \in [0,1]$, T_t is a linear operator and G is a lattice-subspace of E. Consider $(u_n)_n \subset G$ the sequence of continuous functions on $[0,1]$ defined by

$$u_n(t) = 1 - nt \text{ for all } n \in \mathbb{N}.$$

Now, we define the sequences $(x_n)_n \subset G$ and $(y_n)_n \subset E$ by:

$$x_n(t) = u_n(t) \vee_G 0, \text{ and}$$
$$y_n(t) = u_n(t) \vee 0 \text{ for all } n \in \mathbb{N} \text{ and } t \in [0,1].$$

We will calculate

$$\bigwedge_{n \in \mathbb{N}}{}_G x_n \quad \text{and} \quad \bigwedge_{n \in \mathbb{N}} y_n$$

and then

$$T_0\left(\bigwedge_{n \in \mathbb{N}}{}_G x_n\right), \quad \bigwedge_{n \in \mathbb{N}} T_0(x_n)$$

$$T_0\left(\bigwedge_{n \in \mathbb{N}} y_n\right), \quad \bigwedge_{n \in \mathbb{N}} T_0(y_n).$$

I) $x_0(t) = 1$

$\quad x_1(t) = (1-t) \vee_G 0 = 1 - t$

$\quad x_2(t) = (1-2t) \vee_G 0 = 1 - t$

$\quad x_3(t) = (1-3t) \vee_G 0 = 1 - t$

$\quad$ ....

55

It follows that

$$\left(\bigwedge_{n\geq 1}{}_G x_n \right)(t) = 1 - t$$

and hence

$$T_t\left(\bigwedge_{n\geq 1}{}_G x_n \right) = 1 - t \text{ for all } n \in \mathbb{N}.$$

We have $T_0\left(\bigwedge_{n\geq 1}{}_G x_n \right) = 1$, $T_0(x_n) = 1$, and hence $\bigwedge_{n\geq 1} T_0(x_n) = 1$.

It follows that :

$$T_0\left(\bigwedge_{n\geq 1}{}_G x_n \right) = \bigwedge_{n\geq 1} T_0(x_n).$$

II) $y_0(t) = 1$

$y_1(t) = (1-t) \vee 0 = 1 - t$

$$y_2(t) = (1-2t) \vee 0 = \begin{cases} 1-2t, 0 \leq t \leq \dfrac{1}{2} \\ 0 \quad, \dfrac{1}{2} < t \leq 1 \end{cases}$$

$$x_3(t) = (1-3t) \vee 0 = \begin{cases} 1-3t, 0 \leq t \leq \dfrac{1}{3} \\ 0 \quad, \dfrac{1}{3} < t \leq 1 \end{cases}$$

....

It follows that

$$\bigwedge_{n\geq 1} y_n = 0$$

and hence

$$T_t\left(\bigwedge_{n\geq 1} y_n \right) = 0 \text{ for all } t \in [0,1].$$

56

Consequently $T_0\left(\bigwedge_{n\geq 1} y_n\right) = 0$.

But $T_0(y_n) = y_n(0) = 1$ for all $n \geq 1$, and therefore $\bigwedge_{n\geq 1} T_0(y_n) = 1$.

It follows that :

$$T_0\left(\bigwedge_{n\geq 1} y_n\right) \neq \bigwedge_{n\geq 1} T_0(y_n).$$

11. EXTENSION OF THE RESTRICTED-LATTICE OPERATORS

Let E be a vector lattice, $G \subseteq E$ a lattice-subspace, F a Dedekind complete vector lattice and $T : G \to F$ a linear operator. Recall that T is a *G-lattice operator* (or a *restricted-lattice operator*) if
$$T\left(u \vee_G v\right) = T(u) \vee T(v) \text{ for all } u, v \in G .$$

Proposition 11.1. *a) Let E be a vector lattice and $M \subset E$ be a wedge (that is for all $x, y \in M$ and $\alpha \geq 0$, we have $x + y \in M$ and $\alpha x \in M$). Then $G = M - M$ is a vector subspace of E .*
b) If M is closed under finite suprema considered in G (that is for all $u, w \in M$, there exists $u \vee_G w \in G$ and $u \vee_G w \in M$), then G is a lattice-subspace of E .

Proof.
a) Let $u_1 - u_2$ and $w_1 - w_2 \in G$ $(u_1, u_2, w_1, w_2 \in M)$. Then
$$(u_1 - u_2) + (w_1 - w_2) = (u_1 + w_1) - (u_2 + w_2) \overset{M \text{ wedge}}{\in} M - M = G .$$

Let $u - v \in G (u, v \in M)$ and $\alpha \in \mathbb{R}$. Then:
Case 1. $\alpha \geq 0 \Rightarrow \alpha(u - v) = \alpha u - \alpha v \overset{M \text{ wedge}}{\in} M - M = G .$
Case 2. $\alpha < 0$. Let $\beta = -\alpha > 0 \Rightarrow \alpha(u - v) = \beta(v - u) \overset{Case 1}{\in} G .$

b) Let $v_1 - w_1 \in G$ and $v_2 - w_2 \in G$ $(v_1, w_1, v_2, w_2 \in M)$.

We use the following identity:
$$(v_1 - w_1) \vee (v_2 - w_2) = (v_1 + w_2) \vee (w_1 + v_2) - (w_1 + w_2) \quad (35)$$

applied for "$\vee_G$" instead of "$\vee$" (see [LZ], Theorem 11.5 (v)).

Then, there exists $(v_1 - w_1) \vee_G (v_2 - w_2)$ in G, and
$$(v_1 - w_1) \vee_G (v_2 - w_2) \overset{(35)}{=} (v_1 + w_2) \vee_G (w_1 + v_2) - (w_1 + w_2) \in M - M = G,$$

since M is an wedge closed under finite suprema.

Hence, G is a lattice-subspace of E.

Proposition 11.2. *Let E be a vector lattice, $M \subset E$ an wedge, F a Dedekind complete ordered vector space and $G = M - M$. Consider $H \subset M$ a majorizing vector subspace in E (that is for all $x \in E$, there exists $v \in H$ such that $x \leq v$) and $T : H \to F$ a positive linear operator. Denote $\overline{T} : E \to F$ the operator defined by $\overline{T}(x) = \inf\limits_{\substack{v \in H \\ v \geq x}} T(v)$.*

a) *Then, the following are equivalent:*
   i) T extends uniquely to a positive linear operator $L : G \to F$ such that
     $L = \overline{T}$ *on M ;*
   ii) $\overline{T}$ additive on M ;
b) *If, moreover, M is closed under finite suprema considered in G, H is a lattice-subspace and $T(z_1 \vee_H z_2) = T(z_1) \vee T(z_2)$, for all $z_1, z_2 \in H$ (hence T is a H-lattice operator), then L is a G-lattice operator.*

Proof. We know that $\overline{T}$ is increasing, sublinear and that $\overline{T} = T$ on H.
a) $i) \Rightarrow ii)$. Obviously.

   $ii) \Rightarrow i)$. We define $L : G \to F$ by $L(v - w) = \overline{T}(v) - \overline{T}(w)$ for all $v, w \in M$ and we remark that this operator has the following properties.

1). *L is well-defined.* Indeed, let $v_1 - w_1 = v_2 - w_2 \in G$ $(v_1, v_2, w_1, w_2 \in M)$

$$v_1 + w_2 = v_2 + w_1 \text{ in } M \text{ } (M \text{ wedge}) \overset{ii)}{\Rightarrow}$$
$$\overline{T}(v_1) + \overline{T}(w_2) = \overline{T}(v_2) + \overline{T}(w_1) \Rightarrow$$
$$L(v_1 - w_1) = L(v_2 - w_2).$$

2). $L \geq 0$. Indeed, let $v - w \geq 0$ in E $(v, w \in M) \Rightarrow v \geq w$

$$\overset{\overline{T} \text{ increasing}}{\Rightarrow} \overline{T}(v) \geq \overline{T}(w) \Rightarrow \overline{T}(v) - \overline{T}(w) \geq 0 \Rightarrow L(v-w) \geq 0.$$

3). L *linear*. It suffices to prove that $L : G \to F$ is additive (because $L \geq 0$ and F is Archimedean and we apply, for example [C1, Proposition 1, p.157]). Let $v_1 - w_1, v_2 - w_2 \in G$ $(v_1, v_2, w_1, w_2 \in M) \Rightarrow$

$$L\big((v_1 - w_1) + (v_2 - w_2)\big) = L\left(\underbrace{(v_1 + v_2)}_{\in M} - \underbrace{(w_1 + w_2)}_{\in M}\right) =$$

$$= \overline{T}(v_1 + v_2) - \overline{T}(w_1 + w_2) \overset{ii)}{=} \overline{T}(v_1) + \overline{T}(v_2) - \overline{T}(w_1) - \overline{T}(w_2) =$$

$$= L(v_1 - w_1) + L(v_2 - w_2).$$

4). $L = \overline{T}$ on M (and, consequently, $L = T$ on H, that is, L extends T). Obviously L extends uniquely T such that $L = \overline{T}$ on M. Because M is closed under finite suprema considered in G, according to Proposition 11.1 it follows that G is a lattice-subspace.

b) Firstly we will prove the following:

Lemma 11.3. $\overline{T}(u \vee_G v) = \overline{T}(u) \vee \overline{T}(v)$ *for all* $u, v \in G$.

Proof of this Lemma.

"$\geq$" Since $u \vee_G v \geq u, v$ and $\overline{T}$ is increasing, it follows that

$$\overline{T}(u \vee_G v) \geq \overline{T}(u), \overline{T}(v)$$

and hence,

$$\overline{T}(u \vee_G v) \geq \overline{T}(u) \vee \overline{T}(v).$$

"$\leq$"

$$\overline{T}(u) \vee \overline{T}(v) = \left(\bigwedge_{\substack{z_1 \in H \\ z_1 \geq u}} T(z_1)\right) \vee \left(\bigwedge_{\substack{z_2 \in H \\ z_2 \geq v}} T(z_2)\right) \overset{\text{the distributive laws}}{=}$$

$$= \bigwedge_{\substack{z_1 \in H, z_2 \in H \\ z_1 \geq u, z_2 \geq v}} \big(T(z_1) \vee T(z_2)\big).$$

(36)

But $z_1 \geq u$, $z_2 \geq v$ with $z_1, z_2 \in H$ and $u, v \in G \Rightarrow$

$$z_1 \vee_H z_2 \in H \ \left(H \text{ being a lattice-subspace}\right)$$

$$\left.\begin{array}{l} z_1 \vee_H z_2 \geq z_1 \geq u \\ z_1 \vee_H z_2 \geq z_2 \geq v \end{array}\right|_{H \subseteq G} \Rightarrow z_1 \vee_H z_2 \geq u \vee_G v .$$

Hence $z_1 \vee_H z_2 \in \left\{ z \in H \,\middle|\, z \geq u \vee_G v \right\} \Rightarrow$

$$\overline{T}\left(u \vee_G v\right) = \inf_{\substack{z \in H \\ z \geq u \vee_G v}} T(z) \leq T\left(z_1 \vee_H z_2\right) \overset{\text{hypothesis of b)}}{=}$$

$$= T(z_1) \vee T(z_2) \text{ for all } z_1, z_2 \in H \text{ such that } z_1 \geq u, z_2 \geq v$$

$$\Rightarrow \overline{T}\left(u \vee_G v\right) \leq \bigwedge_{\substack{z_1, z_2 \in H \\ z_1 \geq u, z_2 \geq v}} \left(T(z_1) \vee T(z_2)\right) \qquad (37)$$

Applying (36) and (37) it follows that $\overline{T}\left(u \vee_G v\right) \leq \overline{T}(u) \vee \overline{T}(v)$, and so the Lemma 11.3 was proved.

Now we return to the *proof of Proposition* 11.2.
We have to prove that $L\left(u \vee_G v\right) = L(u) \vee L(v)$ for all $u, v \in G$.

Let $u, v \in G$, $u = v_1 - w_1$, $v = v_2 - w_2$ $\left(v_1, v_2, w_1, w_2 \in M\right)$.

$$L\left(u \vee_G v\right) = L\left(\left(v_1 - w_1\right) \vee_G \left(v_2 - w_2\right)\right) \overset{\substack{\text{identity (35)} \\ M \text{ wedge}}}{=}$$

$$= L\left(\underbrace{\left(v_1 + w_2\right)}_{\in M} \vee_G \underbrace{\left(w_1 + v_2\right)}_{\in M} - \underbrace{\left(w_1 + w_2\right)}_{\in M} \right)$$

Using that M is closed under finite suprema considered in G and that L is linear, we obtain that

$$L\left(u \vee_G v\right) = L\left(\left(v_1 + w_2\right) \vee_G \left(w_1 + v_2\right)\right) - L\left(w_1 + w_2\right).$$

But $L = \overline{T}$ on M and hence for all $z_1, z_2 \in M$,

$$L\left(z_1 \vee_G z_2\right) = \overline{T}\left(z_1 \vee_G z_2\right) \overset{\text{Lemma 11.3}}{=} \overline{T}(z_1) \vee \overline{T}(z_2) = L(z_1) \vee L(z_2).$$

For $z_1 = v_1 + w_2$ and $z_2 = w_1 + v_2$ it follows that

$$L\big((v_1 + w_2) \vee_G (w_1 + v_2)\big) = L(v_1 + w_2) \vee L(w_1 + v_2).$$

Then, using that L is linear and the equality (35) applied in the vector lattice F, we obtain

$$L(u \vee_G v) = \big(L(v_1) + L(w_2)\big) \vee \big(L(w_1) + L(v_2)\big) - L(w_1) - L(w_2) =$$

$$\stackrel{(35)}{=} \big(L(v_1) - L(w_1)\big) \vee \big(L(v_2) - L(w_2)\big) \stackrel{L \text{ linear}}{=}$$

$$= L(v_1 - w_1) \vee L(v_2 - w_2) =$$

$$= L(u) \vee L(v). \qquad \qquad \square$$

We notice that the idea of this proof was suggested by a proof of Z. Lipecki, for Riesz homomorphisms - see [L2].

In the following section we will recall some examples (known from the literature) for the basic notions explored in this paper.

12. SOME EXAMPLES

VECTOR LATTICES

According to [AB1], many familiar spaces are vector lattices as the following examples show.

Example 1. (*Vector lattice* - see [W, Example 1.3])
The first example of a vector lattice is of course the reals with all the usual operations. Also, $\mathbb{R}^n$ with the usual order is a vector lattice. Recall that the *usual* or *standard order* on $\mathbb{R}^n$ is that in which $(x_1, x_2, ..., x_n) \leq (y_1, y_2, ..., y_n)$ means that $x_k \leq y_k$ for $k = 1, 2, ..., n$. In this order $(x_k) \vee (y_k) = (x_k \vee y_k)$ and $(x_k) \wedge (y_k) = (x_k \wedge y_k)$. Hence $(x_k)^+ = (x_k^+)$, $(x_k)^- = (x_k^-)$ and $|(x_k)| = (|x_k|)$.

There is *another order* on, for example, $\mathbb{R}^2$ which has been studied, namely the *lexicographic* or *dictionary order*. Under this order $(x_1, x_2) \leq (y_1, y_2)$ means that either $x_1 < y_1$ or $x_1 = y_1$ and $x_2 \leq y_2$.

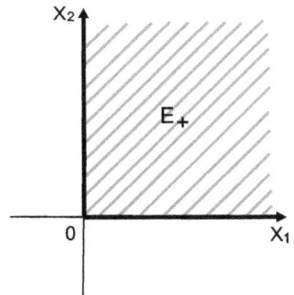 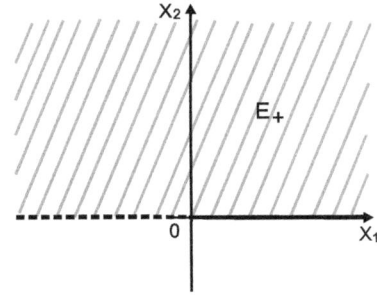

Positive cone E_+ *for the*
usual order on $E = \mathbb{R}^2$

Positive cone E_+ *for the*
lexicographic order on $E = \mathbb{R}^2$

Unlike the usual order on $\mathbb{R}^n$, the lexicographic order fails to have the *Archimedean property* which states that if $nx \leq y$ for all $n \in \mathbb{N}$, then $x \leq 0$. For example $n(0,1) \leq (1,0)$ for all $n \in \mathbb{N}$ yet $(0,1) \nleq (0,0)$. A formulation that is equivalent to a vector lattice being Archimedean is that any infinite affine line in the space (that is, not necessarily going through the origin) has its intersection with the positive cone, closed in the usual topology of the line.

Example 2. (*Archimedean vector lattices* - see [W, Example 1.4]).
Function spaces are important examples of *Archimedean vector lattices*. Let X be a nonempty set and take E to be the space $F(X)$ of all real-valued functions on X. Ordered with the *pointwise order* (under which $f \leq g \Leftrightarrow f(x) \leq g(x)$ for all $x \in X$) and endowed with the pointwise vector operations, E becomes an ordered vector space. Obviously, if f and g are real-valued functions on X, then so are the functions $\Phi(x) = f(x) \vee g(x)$ and $\Psi(x) = f(x) \wedge g(x)$ for all $x \in X$ and it is clear that $\Phi = f \vee g$ and that $\Psi = f \wedge g$, so that E is a *vector lattice*.

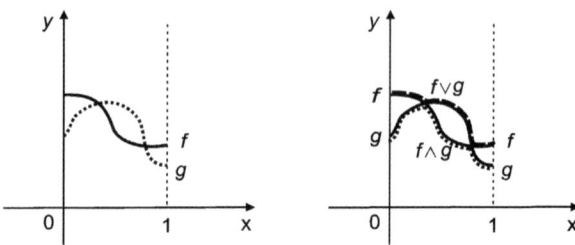

E is *Archimedean* as if $nf \leq g$ for all $n \in \mathbb{N}$, then $nf(x) \leq g(x)$ for all $x \in X$. As $\mathbb{R}$ is Archimedean, it follows that $f(x) \leq 0$ for all $x \in X$ and hence $f \leq 0$ (where 0 is the zero function on X).

E has many vector subspaces which are also vector lattices under the same order, for example the *bounded functions*; if X has a topology then we could take the *continuous functions* or *continuous bounded functions*.

Example 3. (*an ordered vector space which is **not** a vector lattice* - see [W, Example 1.4])
There are many ordered vector spaces which are not vector lattices. The space of *polynomial functions* on $[-1,1]$ with the pointwise vector and order structures is an ordered vector space but not a vector lattice. What could the positive part of the identity function on $[-1,1]$ be? It clearly isn't the pointwise supremum which isn't a polynomial. It takes a little thought and effort to prove that there is no positive part.

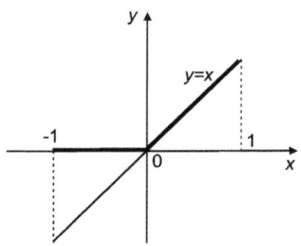

The same is true for the space of differentiable real-valued functions on an interval.

Example 4. (*another ordered vector space which is **not** vector lattice* - see [W, Example 1.4 and pictures]

Let $E = \{f \in C([-1,1]) | \, 2f(0) = f(-1) + f(1)\}$.

If $i(x) = x$ then no matter what we thought i^+ was, we will have:

$$2i^+(0) = i^+(-1) + i^+(1) \geq 0 + i(1) = 1$$

so that

$$i^+(0) \geq \frac{1}{2} > 0.$$

We can now decrease this function slightly on intervals $(-\varepsilon_1, 0)$ and $(0, \varepsilon_2)$ in such a way that it remains an upper bound for both i and the zero function, provided that ε_1 and ε_2 are sufficiently small positive reals.

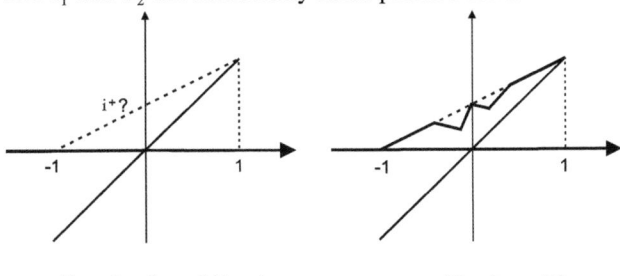

<table>
<tr><td>Can the dotted line be
the graph of i^+ ?</td><td>No, it can't!</td></tr>
</table>

SUBLATTICES

Example 5. (*an ordered subspace which is **not** a sublattice* - see [W, Example 1.4])

A vector subspace of a vector lattice inherits an order from that on the vector lattice. It may or may not be a lattice in that order (that is a sublattice!) For example:

- the *polynomials* on $[-1,1]$ are not a vector lattice even though they sit inside the vector lattice of all continuous real-valued functions on $[-1,1]$.

- the *linear functions* on $[-1,1]$ form a (two dimensional) lattice. However the lattice operations in that subspace are not the same as the lattice operations in whole of $C([-1,1])$, for example, the positive part of i in this subspace is the linear function $x \mapsto \dfrac{x+1}{2}$ (see the first picture on this page) rather than the pointwise supremum. Although this space is a lattice for the inherited order, it fails to be a sublattice. It is a lattice-subspace of $C([-1,1])$.

Example 6. (*sublattices* - see [W, Example 1.6])
We consider the following spaces of sequences: $c_0 = \{(x_n)_n \subset \mathbb{R} | x_n \to 0\}$, the space of all null real sequences, and $c = \{(x_n)_n \subset \mathbb{R} | \exists l \in \mathbb{R} \text{ such that } x_n \to l\}$, the space of all convergent real sequences.
Both c_0 and c are *vector sublattices* of l_∞, when that has been given the pointwise vector and order operations. (Recall that l_∞ is the space of all bounded sequences.)
If $\mathbb{R}^{[0,1]}$, also denoted by $F([0,1])$ is the vector lattice of all real functions defined on $X = [0,1]$ - see Example 2 - then its subspace $C([0,1])$ of all continuous functions is a sublattice ([C1, p.12]). Another sublattice of $C([0,1])$ is its subspace $C_b([0,1])$ of all continuous bounded real functions on $[0,1]$ - see [AB1, p.314]. In turn, $C([0,1])$ is a sublattice of $B([0,1])$, the function space of bounded functions on $[0,1]$ ([AB1, p.321]). Of course, $B([0,1])$ is a sublattice of $F([0,1])$.

LATTICE-SUBSPACES

Example 7. ($\sup(f,g)$ and $\sup_G(f,g)$: *lattice-subspace*)
Let $E = C([0,1])$ with the pointwise algebraic and order structures. Let G be the ordered subspace of E containing all *first-degree polynomial functions* on $[0,1]$. Let $f, g \in G$ defined by $f(t) = t$ and

$g(t) = 1 - t$.

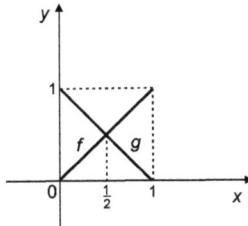

Then we have:

$$\sup(f,g) = \begin{cases} 1-t &, 0 \le t \le \dfrac{1}{2} \\[2mm] t &, \dfrac{1}{2} < t \le 1 \end{cases}$$

$$\inf(f,g) = \begin{cases} t &, 0 \le t \le \dfrac{1}{2} \\[2mm] 1-t &, \dfrac{1}{2} < t \le 1 \end{cases}$$

$$\sup_G (f,g) = 1$$
$$\inf_G (f,g) = 0.$$

Obviously, we have:
$$\inf_G (f,g) \le \inf (f,g) \le \sup(f,g) \le \sup_G (f,g)$$

Remark that:
$$\inf(f,g), \sup(f,g) \in E \backslash G$$
$$\inf_G (f,g), \sup_G (f,g) \in G$$

Hence G is a *lattice-subspace* of E, but G is not a *sublattice*.

SUBLATTICES and IDEALS

Example 8. (*sublattices which are **not** ideals*) As we have noted before, after the observation following Proposition 5.5, $C([0,1])$ is a sublattice of $F([0,1])$

(see Example 6 above), but it is not an ideal. Also (c), which is a sublattice of l_∞ (see Example 6) is not an ideal of l_∞ (see [M-N, p.12]).

IDEALS

Example 9. *(ideals)* The space c_0 is a sublattice of l_∞ (see Example 6), but it is even an ideal of l_∞. (Unlike c_0, the space c of all convergent real sequences is not an ideal, because taking $x_n = 1$ and $y_n = (-1)^n$ for all $n \in \mathbb{N}$ it follows that $\left|(y_n)_n\right| = (x_n)_n \in c$ but $(y_n)_n \notin c$ - see [W, Example 1.8])

Also, the l_p-spaces $0 < p < \infty$ are ideals in the vector lattice $\mathbb{R}^\mathbb{N}$ (also denoted by (s)) of all real sequences - see [AB1, p.321]. (Recall that the usual ordering on $\mathbb{R}^\mathbb{N}$ is the pointwise ordering: $x \geq y$ if and only if $x_n \geq y_n$, for all $n \geq 1$, where $x = (x_n)_n$ and $y = (y_n)_n$. Also, l_p, with $0 < p < \infty$, is the subspace of $\mathbb{R}^\mathbb{N}$ consisting in all $x = (x_n)_n$ such that $\left(\sum |x_n|^p\right)^{\frac{1}{p}} < \infty$.)

RIESZ HOMOMORPHISMS

Example 10. *(Riesz homomorphisms)*
I). Let Ω be a compact space and the Banach lattice $E = C(\Omega)$ (E is endowed with the usual ordering and norm). For all $t \in \Omega$, the functional $f_t : E \to \mathbb{R}$, defined by $f_t = \delta_t$, where $\delta_t(x) = x(t)$, for all $x \in E$ (the *Dirac measure*) is a Riesz homomorphism.

II). Any *projector* P on a vector lattice E is a *Riesz homomorphism*.
(Recall that $Q \subseteq E$ is called a *component* of E, if any $x \in X$ can be written $x = x' + x''$, $x' \in Q$ and $x'' \in Q^\perp = \{z \in E \mid |x| \wedge |y| = 0, \forall y \in Q\}$ and the element x' is denoted by $[Q](x)$; the operator $P : E \to Q$ defined by $P(x) = [Q](x)$ is called a *projector*.)

Proof . According to [D4, Proposition 1 i) $\Leftrightarrow$ iii)] we be shown that if $x_1 \wedge x_2 = 0$ in E, then $P(x_1) \wedge P(x_2) = 0$. But this is immediate since

$0 \le P(x_1) \le x_1$ and $0 \le P(x_2) \le x_2$.   □

III). Let E be a vector lattice, $F = E$ and $\lambda \in \mathbb{R}_+$. Then $T_\lambda : E \to F$ defined by $T_\lambda(x) = \lambda x$ for any $x \in E$ is a *Riesz homomorphism*.

IV). Let E be a vector lattice of functions. For any positive function $x_0 \in E$ such that $x_0 \cdot E \subseteq E$, the operator $T_{x_0} : E \to E$ defined by $T_{x_0}(x) = x_0 \cdot x$, $x \in E$ is a *Riesz homomorphism*, too.

V). [*The embedding in the bidual*]
Let E be a vector lattice and $G \subseteq E'$ an ideal, that is, if $y \in G$ and $x \in E'$ with $|x| \le |y|$, then $x \in G$. It is known that $E' = E^\#$ - see, for example, [C1, p.167], where E' is the regular dual of E and $E^\#$ is the collection of all order bounded functionals on E). Then the canonical map $T : E \to G'$ (defined by $T(x)(f) = f(x)$, for all $x \in E$ and $f \in G$) is a *Riesz homomorphism*.

Proof. According to [D4, P1, i) $\Rightarrow$ v)] we can prove that
$$\left(T(x)\right)^+ = T\left(x^+\right), \forall x \in E.$$
But if $f \in G$:
$$T\left(x^+\right)(f) = \sup\{T(x)(g) | g \in G, 0 \le g \le f\} =$$
$$= \sup\{T(x)(g) | g \in E', 0 \le g \le f\},$$
since G is an ideal in E'.
But then:
$$T\left(x^+\right)(f) = \sup\{g(x) | g \in E', 0 \le g \le f\} =$$
$$= f\left(x^+\right) = T\left(x^+\right)(f) \text{ for all } f \in G.$$ □

VI). ([L3, Lemma]). Let Ω be an (abstract) set and let $\mathfrak{M}$ be an algebra of subsets of Ω. If $E = B(\mathfrak{M})$ is the vector lattice of real-valued bounded functions on Ω which are uniform limits of $\mathfrak{M}$-simple functions, F is an order complete vector lattice (over $\mathbb{R}$) and $\mu : \mathfrak{M} \to F$ is a quasimeasure (i.e

μ is nonnegative and additive), then the positive linear operator $T : B(\mathfrak{M}) \to F$ which extends uniquely μ (according, for example to [B1, Section 3]) is a *Riesz homomorphism* if

$$\mu(M_1 \cap M_2) = \mu(M_1) \wedge \mu(M_2)$$

for all $M_1, M_2 \in \mathfrak{M}$.

Recall that T is the *integral with respect to* μ. It is defined by the following
- for simple functions:

$$T\left(\sum_{i=1}^{n} t_i 1_{M_i}\right) = \sum_{i=1}^{n} t_i \mu(M_i), \, t_i \in \mathbb{R}, \, M_i \in \mathfrak{M} \text{ for all } i = \overline{1,n}, n \in \mathbb{N}^*$$

- for $x \in B(\mathfrak{M})$:

$$T(x) = \sup\{T(s) | s \leq x, \, s \text{ is } \mathfrak{M}\text{ - simple}\} \left(= \int x d\mu\right) \quad.$$

Proof. According to [D4, P1, i) $\Leftrightarrow$ vi)], we need to show that $T(|x|) = |T(x)|$, for all $x \in E$. In case x is $\mathfrak{M}$ - simple, this is immediate since $\mu(M_1) \wedge \mu(M_2) = 0$, whenever $M_1, M_2 \in \mathfrak{M}$ are disjoint. To deduce the general case, it is sufficient to note that for every $\mathfrak{M}$ - simple function s, we have:

$$T(|x|) \leq T(|s|) + T(|x-s|) = |T(s)| + T(|x-s|) \leq$$
$$\leq |T(x)| + 2T(|x-s|)$$

(according to [L1, proof of Theorem 2(a)]). $\qquad\qquad\qquad\square$

VII). ([S, p.60]) Recall that a *normal topological space* is a topological space Ω such that every two disjoint closed sets A, B of Ω have disjoint open neighborhoods U, V. (Recall that the set U is a *neighborhood* for A if and only if U includes an open set U_1 containing A; it follows that U is a neighborhood for A if and only if U it is a neighborhood of all points in A; or, equivalently, U is a neighborhood of A if and only if A is a subset of the interior of U .)

Also recall that a *nowhere dense* set in a topological space Ω is a subset Ω_0 of Ω, whose closure has empty interior. Every subset of a nowhere dense set is nowhere dense, and the union of finitely many nowhere dense sets is nowhere dense. The union of countably many nowhere dense sets, however, need not be nowhere dense .

If E is a vector lattice of real functions on a nonempty set Ω under its canonical order and $F = \mathbb{R}^{\Omega_0}$ (canonical ordered), where Ω_0 is a nonempty subset of Ω, then the restriction map $T : E \to F$ defined by $T(x) = x|_{\Omega_0} \ (x \in E)$ is a *Riesz homomorphism*. If $\Omega_0 = \Omega$ then T is a *normal Riesz homomorphism*. If Ω is a normal topological space, Ω_0 is a closed nowhere dense subset of Ω and $E = C(\Omega)$ is the vector lattice of all continuous real functions on Ω, then T is not a normal Riesz homomorphism.

Proof. Indeed, consider the family $(x_\delta)_{\delta \in \Delta}$ of all functions $x \in E$ satisfying $x|_{\Omega_0} = \{1\}$. This family is directed downwards and order converges to 0 in E, while $T(x_\delta) \downarrow 1$. $\qquad \square$

To draw a conclusion of this example we can say that the restriction of a normal Riesz homomorphism $T : E \to F$ to a vector sublattice of E need not be normal Riesz homomorphism.

References

[AAP] Abramovich, Y.A.; Aliprantis, C.D.; Polyrakis, I.A.: *Lattice-subspaces and positive projections*, Proc. Roy. Irish Acad. **94A**(1994), no.2, 237-253.

[AB1] Aliprantis, C.D.; Border, K.C.: *Infinite Dimensional Analysis, A Hitchhikers's Guide*, Springer Verlag, Berlin, Heidelberg, 1999.

[AB2] Aliprantis, C.D.; Burkinshaw, O.: *Locally solid Riesz spaces*, Academic Press, New York - San Francisco-London, 1978.

[AT] Aliprantis, C.D.; Tourky, R.: *Cones and Duality,* Graduate Studies in Mathematics, Volume **84**, American Mathematical Society, Providence, Rhode Island, 2007.

[B1] Berz, E.: *Verallgemeinerungen eines Satzes von F. Riesz*, Manuscripta Math. **2** (1970), 285-299.

[C1] Cristescu, R.: *Ordered Vector Spaces and Linear Operators*, Ed. Acad. Bucharest, Romania - Abacus Press, Tunbridge Wells, Kent, England, 1976.

[C2] Cristescu, R.: *Ordered structures in normed linear spaces,* Ed. St. Encicl. Buc., 1983 (in Romanian).

[D1] Dăneţ, R–M.: *Geometric and algebraic interpretation of lattice operations* (in Romanian), Proceedings of 7-th Workshop of Department of Mathematics and Computer Science, Technical University of Civil Engineering, Bucharest, Romania 24 May, 2003, Ed. Conspress, 23-25.

[D2] Dăneţ, R.–M.: *How to introduce some basic notions for an order relation. Pictures and new proofs*, 3rd Conference on the History of Mathematics and Teaching of Mathematics, Univ. of Miskolc, May 20-23, 2004.

[D3] Dăneţ, R.–M.: *New formulations of some notions of vector lattice theory* (in Romanian), Proceedings of 8-th Workshop of Department of Mathematics and Computer Science, Technical University of Civil Engineering, Bucharest, Romania 21 May, 2005, Ed. Matrix Rom, 35-38.

71

[D4] Dăneţ, R.–M.: *Riesz homomorphisms. Quasi Riesz homomorphisms,* Order Structures in Functional Analysis, Ed. Acad. Rom., **4**(2001), 45-89.

[D5] Dăneţ, R.–M.: *On some vector lattice concepts,* The Thirteenth Conference of Department of Mathematics and Computer Science, Technical University of Civil Engineering of Bucharest, May 25, 2013, Bucharest, Romania.

[F] Freudenthal, H.: *Teilweise geordnete Moduln.* Proc. Acad. Sci. Amsterdam, **39** (1936), 641-651.

[J] Jameson, G.: *Ordered linear spaces,* Lecture Notes in Math. Springer, 1970.

[K1] Kantorovitch, L.V.: *Sur les espaces semi-ordonnées linéaires et leur applications à la théorie des opérations linéaires,* C.R. Acad. Sci. de l'U.R.S.S. (DAN), **4** (1935), 11-14.

[K2] Kantorovitch, L.V.: *Sur les propriétés des espaces semi-ordonnées linéaires,* C.R. Acad. Sci. de l'U.R.S.S. (DAN), **202** (1936), 813-816.

[L1] Lipecki, Z.: *Extensions of positive operators and extreme points-II,* Colloq. Math., **42** (1979), 285-289.

[L2] Lipecki, Z.: *Extensions of vector lattices revisited,* Indag. Math. **47** (1985), 229-233.

[L3] Lipecki, Z.: *Riesz type representation theorems for positive operators,* Math. Nachr. **131** (1987), 351-356.

[LZ] Luxemburg, W.A.J.; Zaanen, A.C.: *Riesz Spaces I,* North Holland, Amsterdam, London, 1971.

[M-N] Meyer-Niebery, P.: *Banach Lattice,* Springer-Verlag, Berlin Heidelberg, 1991.

[M] Miyajima, S.: *Structure of Banach quasi-sublattices,* Hokkaido Math. J. **11**(1983), 83-91.

[P1] Polyrakis, I.A.: *Lattice Banach Spaces order-isomorphic to l_1*, Math. *Proc. Cambridge Phil. Soc.,* **34**(1983), 519-522.

[P2] Polyrakis, I.A.: *Finite-dimensional lattice-subspaces of $C(\Omega)$ and curves of $\mathbb{R}^n$,* Transactions of the American Mathematical Society, Volume **348**, Number **7**, July 1996.

[P3] Polyrakis, I.A.: *Minimal lattice-subspaces*, Transactions of the American Mathematical Society, Volume **351**, Number **10**, pages 4183-4203, S 0002-9947(99)02384-3. Article electronically published on April 20, 1999.

[P4] Polyrakis, I.A.: *Lattice-subspaces and Positive Bases in Function Spaces*, Positivity 7, 267-284, 2003, Kluwer Academic Publishers. Printed in Netherlands.

[R] Riesz, F.: *Sur la décomposition des opérations fonctionnelles linéaires.* Atti Congr. Internaz. Mat., Bologna 1928, **3** (1930) 143-148; Oeuvres Complètes II, 1097-1102, Budapest, 1960 (§ 83).

[S] Schaefer, H.H.: *Banach lattices and positive operators*, Springer-Verlag, Berlin, Heidelberg, New York, 1974.

[W] Wickstead, A.W.: *Vector and Banach lattices*, Queen's Univ Belfast, expository talk workshop Positivity V, July 2007, available at <www.qub.ac.uk/puremaths/Staff/Anthony Wickstead/Positivity V/Positivity V.html>

INDEX